Holger Schütz, Peter M. Wiedemann,
Wilfried Hennings, Johannes Mertens,
and Martin Clauberg
Comparative Risk Assessment

Related Titles

Bhagwati, K.
Safety Management
For Executives and Managers

2007
ISBN 3-527-31583-7

Tanzil, D.
Transforming Sustainability Strategy into Action
The Chemical Industry

2005
ISBN 0-471-64445-5

Ericson, C. A.
Hazard Analysis Techniques for System Safety

2005
ISBN 0-471-72019-4

Aven, T.
Foundations of Risk Analysis
A Knowledge and Decision-Oriented Perspective

2003
ISBN 0-471-49548-4

Paustenbach, D. J. (ed.)
Human and Ecological Risk Assessment
Theory and Practice

2002
ISBN 0-471-14747-8

*Holger Schütz, Peter M. Wiedemann,
Wilfried Hennings, Johannes Mertens,
and Martin Clauberg*

Comparative Risk Assessment

Concepts, Problems and Applications

WILEY-VCH Verlag GmbH & Co. KGaA

The Authors

Holger Schütz
Forschungszentrum Jülich GmbH
Mensch, Umwelt, Technik
52425 Jülich

Dr. Peter M. Wiedemann
Forschungszentrum Jülich
Mensch, Umwelt, Technik
52425 Jülich

Dipl.-Ing. Wilfried Hennings
Foschungszentrum Jülich
Mensch, Umwelt, Technik
52425 Jülich

Dr.rer.nat. Johannes Mertens
Forschungszentrum Jülich
Mensch, Umwelt, Technik
52425 Jülich

Dr. Martin Clauberg
University of Tennessee
1060 Commerce Park
Oak Ridge, TN 37830
USA

■ All books published by Wiley-VCH are carefully produced. Nevertheless, authors, editors, and publisher do not warrant the information contained in these books, including this book, to be free of errors. Readers are advised to keep in mind that statements, data, illustrations, procedural details or other items may inadvertently be inaccurate.

Library of Congress Card No.:
applied for

British Library Cataloguing-in-Publication Data:
A catalogue record for this book is available from the British Library.

**Bibliographic information published by
Die Deutsche Bibliothek**
Die Deutsche Bibliothek lists this publication in the Deutsche Nationalbibliografie; detailed bibliographic data are available in the Internet at <http://dnb.ddb.de>.

© 2006 WILEY-VCH Verlag GmbH & Co. KGaA, Weinheim

All rights reserved (including those of translation in other languages). No part of this book may be reproduced in any form – by photoprinting, microfilm, or any other means – nor transmitted or translated into machine language without written permission from the publishers. Registered names, trademarks, etc. used in this book, even when not specifically marked as such, are not to be considered unprotected by law.

Cover Design   Adam Design, Weinheim
Typesetting   TypoDesign Hecker GmbH, Leimen
Printing   betz-druck GmbH, Darmstadt
Binding   Litges & Dopf GmbH, Heppenheim

Printed in the Federal Republic of Germany
Printed on acid-free paper

ISBN-13   978-3-527-31667-0
ISBN-10   3-527-31667-1

Table of Contents

Foreword *IX*

Acknowledgements *XI*

Summary *1*

1	**Introduction** *7*	
2	**Concepts and Definitions** *13*	
2.1	Risk *13*	
2.2	Danger/Potential for Damage–Hazard *16*	
2.3	Incertitude and Uncertainty *17*	
2.4	Risk Assessment *17*	
2.5	Risk Evaluation *18*	
2.6	Risk Comparison *20*	
2.7	Risk Management *21*	
2.8	Risk Regulation *23*	
2.9	Risk Communication *25*	
3	**CRA in Practice** *27*	
3.1	Political Environment of CRA *27*	
3.1.1	CRA in the USA *27*	
3.1.2	CRA in Europe *29*	
3.1.3	CRA in Germany *32*	
3.2	Project Case Studies *34*	
3.2.1	US EPA Unfinished Business Project *34*	
3.2.2	Comparative Risk Framework Methodology and Case Study (US EPA) *40*	
3.2.3	Global Burden of Disease (GBD) Study *46*	
3.2.4	ExternE Project *55*	
3.2.5	Comprehensive Assessment of Energy Systems Project *66*	
3.2.6	Classification of Carcinogenic Airborne Pollutants for the German TA Air Novella *75*	
3.2.7	Summary *75*	

Comparative Risk Assessment. Holger Schütz, Peter M. Wiedemann,
Wilfried Hennings, Johannes Mertens, and Martin Clauberg
Copyright © 2006 WILEY-VCH Verlag GmbH & Co. KGaA, Weinheim
ISBN 3-527-31667-1

4	**The Empirical Foundations of CRA** *77*	
4.1	A Starting Point for Risk Comparisons *77*	
4.2	Risk Comparisons as a Means of Risk Communication *78*	
4.3	Procedural Challenges *83*	
4.3.1	Framework: What is to be Kept in Mind when Determining the Systems Limit Options for CRA? *83*	
4.3.2	Risk Categories: What Influence Do They Have on Comparisons? *85*	
4.3.3	Attributes: What Influence Do They Have on Comparisons? *88*	
4.3.4	Assessment of the Attributes: What Influence Does the Measure of a Risk Have on Comparisons? *90*	
4.3.5	Comparisons: What Influence Does the Nature of the Comparison Have on the Comparison? *95*	
4.4	CRA Negotiations Under Conflict *97*	
4.4.1	Pitfalls in the Evaluation of One's Own Position and Interests *98*	
4.4.2	Pitfalls in the Evaluation of Other Parties *99*	
4.4.3	Pitfalls in the Evaluation of Suggestions *101*	
4.4.4	Pitfalls in the Evaluation of Outcomes *102*	
5	**Conceptual Framework for an Integrated Comparative Risk Evaluation** *105*	
5.1	Methodological Problems of a CRA *105*	
5.1.1	Problem: Uncertainty and Incertitude *105*	
5.1.2	Problem: Evaluative Criteria *109*	
5.1.3	Consequences for the Conduct of a CRA *112*	
5.2	Methodology of Comparative Risk Evaluation *114*	
5.2.1	Multiattribute Risk Evaluation: Quantitative and Qualitative Approaches *115*	
5.2.2	Conduct of a Multiattribute Evaluation *117*	
6	**The Practical Implementation of CRA** *135*	
6.1	Limits of Comparability *135*	
6.2	Goals of a Comparative Evaluation of Risks *138*	
6.2.1	Preliminary Analysis *138*	
6.2.2	Risk Assessment *141*	
6.2.3	Risk Evaluation *143*	
6.2.4	Risk Management *144*	
6.3	Participation Models *145*	
6.4	CRA Participants *146*	
6.5	The Sequence of Events in a CRA *148*	
6.6	The Organization of Communication *150*	

Appendix 1: Risk Metrics *155*

Appendix 2: Multiattribute Comparative Risk Evaluation (MCRA) *165*

Appendix 3: Comparative Evaluation of Unclear Risks *181*

Appendix 4: Comparative Evaluation of the Risks of Hazardous Incidents *189*

References *195*

Abbreviations of Organizations *213*

Index *215*

Foreword

The International Risk Governance Council (IRGC) is delighted to have been able to support the Forschungszentrum Jülich GmbH's Man, Environment and Technology Programme Group by funding the translation of their original 2004 text into English. We very much hope that having the report available in English will help to increase worldwide readership of this contribution to the understanding of risk and its governance.

The IRGC is a private, independent, not-for-profit foundation based in Geneva, Switzerland. The IRGC was founded in 2003. Its mission is to support governments, industry, NGOs, and other organizations in their efforts to deal with major and global risks facing society and to foster public confidence in risk governance. We work to achieve this mission by reflecting different views and practices and providing independent, authoritative information, by improving the understanding and assessment of risk and the ambiguities involved, by exploring the future of global risk governance, and by designing innovative governance strategies. We focus on issues, whether human-induced or natural, which have international implications and have the potential for harm to human health and safety, the economy, the environment, and/or to the fabric of society at large. We endeavor to work and communicate in ways that account for the needs of both developed and developing countries.

The establishment of the IRGC was the direct result of widespread concern within the public sector, the corporate world, academia, the media, and society at large that the complexity and interdependence of an increasingly large number of risk issues were making it ever more difficult for risk managers to develop and implement adequate risk governance strategies. Consequently, the IRGC is committed to promoting a multidisciplinary, multisectoral, and multiregional approach to risk governance. The subject of comparative risk assessment (CRA) was identified as a priority area for the IRGC from the date of its founding, precisely because it offers a means of improving risk decision-making on a global basis.

All those who make risk-related decisions, whether government ministers, regulators, or company directors, require sound knowledge on which to base their decisions, wherever possible including the best scientific knowledge available. Often, decision-makers are confronted by the need to make decisions for which they must allocate resources to one or more of several different problems, and are required to

Comparative Risk Assessment. Holger Schütz, Peter M. Wiedemann,
Wilfried Hennings, Johannes Mertens, and Martin Clauberg
Copyright © 2006 WILEY-VCH Verlag GmbH & Co. KGaA, Weinheim
ISBN 3-527-31667-1

do so in the absence of any objective means of comparing the risks or the impact of their decisions. CRA offers a possible means of providing a scientific basis for such decisions.

As the authors of this book make clear, there is a need to further study and understand how best CRA can fulfill its full potential. There is a growing body of knowledge, and the publication of this book is a most helpful addition to it. We are pleased to have been able to play a role in helping to bring Forschungszentrum Jülich GmbH's report to a wider audience.

Finally, the IRGC extends its thanks to Simon Milligan for the excellence of his translation of this valuable contribution to the debate about CRA and its place in the total risk governance process.

Pittsburgh PA, USA, April 2006

Professor M Granger Morgan
University and Lord Chair of Engineering and Head of the
Department of Engineering and Public Policy,
Carnegie Mellon University;
Chairman of the IRGC's Scientific and Technical Council

Acknowledgements

This book is a translated and revised version of a report for the German Federal Office for Radiation Protection (BfS), Grant StSch 4217: Schütz, H., Wiedemann, P.M., Hennings, W., Mertens, J. and Clauberg, M. (2004) *Vergleichende Risikobewertung. Konzepte, Probleme, Anwendungsmöglichkeiten* (Reihe Umwelt/Environment 45). Jülich: Forschungszentrum Jülich GmbH.

The translation was prepared by Simon Milligan and funded by the *International Risk Governance Counsil (IRGC)* [http://www.irgc.org/].

Summary

Decisions dealing with environmental health risks must be made under the most widely differing contexts. These range from decisions of the private sphere through regulatory decisions to those on the global political level. A glance at some contemporary problems in society shows that the spectrum of risk-related decisions is very broad:
- Should the use of a particular substance be regulated (e.g., by setting limit values)?
- Can a substance known to be harmful be replaced by another (e.g., by substituting chemicals)?
- Where should the emphasis of health policy be placed (e.g., cancer prevention)?
- Which energy systems should be used (e.g., fossil fuels, renewable energy sources)?
- Which methods of agricultural production should be preferred (e.g., conventional agriculture, genetically modified crops)?

The solutions to such challenges require regulatory and political decisions that should be based on a thorough and rational analysis of risks and decision options. The above problem statements make it clear that *comparative* risk assessments (CRAs) can play an important role in the regulation of risks.

The claim that comparative risk evaluation forms a rational foundation for making environment health-related decisions has led, especially in the USA, to a critical debate in which a range of arguments have been brought against CRA. Some of the criticisms are of a fundamental nature. One such is that risks are frequently so qualitatively heterogeneous, for instance in respect to their origin or the type of damage they cause, that a meaningful comparison cannot be made. Another argument centers on the considerable knowledge gaps of environment health risks and deduces that comparative evaluations of these risks are not possible or, at the least, not meaningful. A third argument sees in CRA an obstacle to the possible participation of the public in decisions on environmental health risks. Relying on CRA for such decisions is seen to create the danger of an "expertocracy" because the risk assessments upon which a CRA must be based are frequently very complex and, thus, can hardly be followed by nonexperts.

Comparative Risk Assessment. Holger Schütz, Peter M. Wiedemann,
Wilfried Hennings, Johannes Mertens, and Martin Clauberg
Copyright © 2006 WILEY-VCH Verlag GmbH & Co. KGaA, Weinheim
ISBN 3-527-31667-1

In the light of this situation, the objectives set for this study were to evaluate existing approaches to CRA, to explore the possibilities of improving such comparisons, to discuss the opportunities for participation, and to design processes that will increase public acceptance of political decision-making as regards environmental health risks.

The following questions are addressed:
- What prior efforts of comparative risk evaluations are there? What gaps and problems do these present?
- What types of comparative risk evaluation can be distinguished, and which of these, within existing administrative constraints, are meaningful and worthy of development?
- How are the individual steps of a comparative risk evaluation to be designed, and what advice can be gleaned from the substantial body of literature on social science risk research?
- What special demands does an integrated comparative risk evaluation have, in particular one that involves a variety of evaluators from science and society?
- What are the chances of success for comparative risk evaluations, and where do their limitations lie?

The results of this study are summarized as a number of recommendations, as discussed below.

1. CRA can draw upon a range of developed and established instruments
Comparative risk evaluation can be conceptualized as a multiattribute evaluation procedure. Thus, it can build upon the approach of multiattribute decision-making, which allows for a theoretically sound and structured progression by way of manageable individual steps. For each single step (structuring the problem, structuring and weighting the attributes, sensitivity analysis, etc.) there is a range of practically tested techniques.

One of the strengths of this approach is that it facilitates an explicit examination of assumptions and values and thus aids in a transparent comparative risk evaluation. This approach is therefore suitable for precisely those CRA processes in which a variety of evaluators – experts and stakeholders – take part. A multitude of highly detailed participation procedures and models exists for organizing and implementing such multistakeholder evaluations.

In addition, as regards the practical organization and implementation of CRAs, a rich store of experience can be called upon. In the USA in particular, more than 100 CRAs have already been implemented for which summaries of the experience gained are available.

2. CRA requires scientific expertise and must be developed further
An important prerequisite for a CRA is the harmonization of methods of risk assessment and evaluation from different fields. To this end, the characterization of risks should be regarded as a separate step. The entire process requires scientific

expertise and needs to be further developed to allow for an efficient implementation.

Risk assessment begins with the identification of hazards. Three problem areas are of significance here: (a) the degree of evidence required to substantiate a causal link between the causes and effects in question, (b) the classification of an effect as adverse or undesirable, and (c) possible exposure of subjects of protection.

The evaluation of evidence is a substantial problem. Here categories ("how strong is the suspicion?") must be developed that can be unambiguously operationalized. Worst case scenarios are, due to their arbitrary nature – it is always possible to imagine even more severe scenarios – no workable basis for the assessment and evaluation of risks. Dose–response assessments should be determined in accordance with standardized and harmonized methods. In the light of the importance of hazards, exposure assessments are also of considerable significance.

The risk characterization thus brings together the results of the identification of hazards, dose–response assessments, and exposure assessments. This examination of the data is also a factual prerequisite for comparative analyses. Hence the risk characterization should be regarded as a special stage in the process.

Risk evaluation constitutes the link between the predominantly scientific/technical risk assessment and a sociopolitically oriented valuation of risks. A consensus on what are tolerable risks, reached through societal debate, can be the basis for an evaluation of quantifiable risks. Many deliberations must cope with (as yet) unquantifiable risks, and thus criteria for differentiating – again on the basis of scientific expertise – between averting a substantiated danger (with unambiguous regulatory requirements, primarily through limit values) and precautionary measures need to be developed. It is, furthermore, undetermined which suspicions of risk are strong enough to justify the application of the precautionary principle as a way of reaching an initial basis for making comparisons with other risks. However, standards of quality neither for individual studies nor for the overall scientific understanding of risk suspicions have yet been developed.

Another problem area is the evaluation of risks involving new or developing technologies (such as nanotechnology or genetic manipulation). Special methods, including prognostic procedures, for the early recognition of potential hazards and their relevance need to be developed. To this end, it should be ensured that the conceptualization of such methods be consistent and provide for the ability to compare different technologies.

3. CRA can provide important information for all stages of risk regulation
In the risk regulation process, different individual steps with their own specific objectives can be differentiated. CRA can provide substantial information at all stages; thus, it is clear that even individual components of a CRA – in the sense of risk-related comparisons – can prove useful for the respective objectives.

The *preliminary analysis* is concerned with the analysis of a novel risk potential, with a new technology's public mobilization potential, and with an initial analysis of the urgency for a risk evaluation. The benefit of a CRA lies here in the comparison of new technology fields, in the comparison of public risk perceptions for different cases, and in the comparison of hazardous substances with regard to their emission data, exposure characteristics, and toxicity.

Risk assessment is focused on the evaluation of evidence. This is where scientific controversy is often found and a comparison of different evidence evaluations, for instance with the use of tried and tested guidelines and categories of evidence, could contribute considerably to the solution of the problem. Of note here is the method of *comparative evaluation of unclear risks* in which similarities and differences that are demonstrable between different experts or groups of experts in the evaluation of evidence of the existence of a threat are compared.

In the regulatory step of *risk evaluation* there are four different opportunities for making risk-related comparisons: (a) the evaluation of a pollutant's potency, (b) the evaluation of exposure to such a pollutant, (c) the evaluation of the vulnerability of populations, and (d) the comparative evaluation of the various risks.

At the center is the opportunity for comparative evaluation of various risks. Particularly noteworthy here is the procedure known as *multiattribute comparative risk evaluation*, in which the evaluative dimensions (attributes) of known risks are compared by one or more evaluators (stakeholders). This essentially follows the classic approaches of CRA. The outcome of such a multiattribute CRA is a ranking of risks on the basis of an (preferably quantitative) assessment of the health consequences and their evaluation.

A number of risk-related comparisons also lend themselves to *risk management*, for example in the selection of technological alternatives, or in the siting search for locations of facilities with potential risks. Cost and benefit aspects are typically included in such comparisons.

4. CRA, as a combination of scientifically based risk assessments and value judgments, requires the cooperation of experts and societal stakeholders

Experts – such as the authors of scientific risk assessments – and the general public frequently have very different understandings and interpretations of risk assessments. One substantial problem, from the point of view of experts, is that the final results of analyses are separated from their principal constraints, methodological uncertainties, and scope, of which the public remains unaware.

What is basic to the understanding and role of risk assessments is, furthermore, the idea of risk itself. It has been shown that the technical conception of experts is, from the public's point of view, extremely narrow and encompasses only a fraction of the aspects and values that the general public – broadly represented by societal stakeholders – consider important to an appraisal of risk. Even the consideration of frequency and loss as equivalent, which is derived from the insurance industry, is – as it is among experts – disputed. Both factors are treated by lay people (i.e., those who are not risk experts) individually; in particular, the upper limit of potential damages is seen as an independent issue and is increasingly demanded.

In addition, the concept of risk underlying risk assessments usually encompasses only a few of the dimensions of loss, often only loss of life and harm to health, and, in rare cases, loss of property. The public mostly looks at some of the other dimensions and concomitant circumstances of risks, such as the timeframe in which harmful effects occur, the physical, i.e., spatial, extent of losses, the unavoidability of risks, evacuations, resettlements, and other conspicuous social aspects. Within the context of specific CRA procedures, experts on the one hand and stakeholders on the other must clarify which aspects should be taken into account within the CRA.

A consensually accepted and successful CRA can, therefore, only be based on the cooperation of experts (in risk assessment and management) and societal stakeholders as representatives of public opinion.

5. CRA requires a risk communication program

An important prerequisite for the success of a CRA project is good communication. What are essential here are the setting of clear objectives, explicit and comprehensible definitions of the risks being compared and of the assessment criteria and units of measure, and a comprehensible characterization of risk.

Every CRA assumes that the parties involved are sufficiently well informed that they are able to deal with the comparative assessment of risks. Beyond this, however, several aspects of the process of communication are to be noted. Chief amongst these is the realization of fairness, competence, and trust.

The following aspects need to be distinguished for the public communication of the results of a CRA: (a) acceptance of CRA methods, (b) the clarification of a CRA's contribution to the understanding of the scale of a risk, and (c) acceptance of the results of a CRA.

Acceptance of a CRA is only possible when it succeeds in creating a mature understanding of risk so that it can be rationally weighed. The clarification of a CRA's contribution to the understanding of a risk depends upon whether and how existing information gaps are filled. The exchange of viewpoints and attitudes plays an important role in this. Such an exchange is also essential for the acceptance of CRA results.

1
Introduction

Decisions dealing with environmental health risks must be made under the most widely differing contexts. These range from decisions of the private sphere through regulatory decisions to those on the global political level. While such risk-related decisions in one's private life (for instance, engaging in dangerous or extreme sports or taking out an accident insurance policy) are mostly made without a precise analysis of risks or of the advantages and disadvantages of the available alternatives, regulatory and political risk-related decisions should be based on a thorough analysis of risks and decision options.

A glance at some contemporary problems in society shows that the spectrum of risk-related decisions is very broad:
- Should the use of a particular substance be regulated (e.g., by setting limit values)?
- Can a substance known to be harmful be replaced by another (e.g., by substituting chemicals)?
- Where should the emphasis of health policy be placed (e.g., cancer prevention, AIDS awareness)?
- Which energy systems should be used (e.g., fossil fuels, renewable energy sources)?
- Which methods of agricultural production should be preferred (e.g., conventional agriculture, genetically modified crops)?

The tasks of risk assessment, risk evaluation, and risk management are, in Germany, administered by different institutions.[1] Thus, the responsibility is distributed among various ministries at the federal level (e.g., BMU, BVEL, BfG) which each consult the expert knowledge of subordinate federal agencies (e.g., BfS, UBA, RKI), private legal entities (e.g., Senatskommission zur Bewertung Maximaler Arbeitsplatzkonzentrationen der DFG), and expert committees (e.g., SSK) which are assigned to them. Numerous committees (e.g., LAI, LAUG) are also involved at the state level. This profusion of competencies and administrative levels may lead to a loss of clarity and confusion. In its final report on the reorganization of the methods and structures for risk evaluation and standard-setting in environmental

---
1) See the German Risk Commission's analysis (2003) for details.

Comparative Risk Assessment. Holger Schütz, Peter M. Wiedemann,
Wilfried Hennings, Johannes Mertens, and Martin Clauberg
Copyright © 2006 WILEY-VCH Verlag GmbH & Co. KGaA, Weinheim
ISBN 3-527-31667-1

health protection of the Federal Republic of Germany, the German Risk Commission came to the conclusion that

> *The result of all this is that the way potential health risks are dealt with is usually determined by a fortuitous or interest-driven perception of the problem on the part of the public or the media. In the recent past a number of conflicts of objectives have made it clear that there is a lack of effective crisis management. Participation by the public and by interest groups in risk regulation tends to be the exception, and it is unsystematic and ponderous. Hesitant and contradictory regulation by the governmental bodies involved, selective information of the parties concerned, and communication deficits combine to produce a situation where on the one hand relatively minor risks occupy an important position in public perception, whereas on the other, serious risks are underestimated or even "swept under the carpet".* (German Risk Commission 2003, p. 18)

This appraisal makes it clear that comparative risk assessments can play an important role in the regulation of risks.

Comparisons of risks are, in and of themselves, nothing new. Indeed, every evaluation of a risk implies some comparison: either a comparison of the risk under consideration with an environmental standard (a threshold, benchmark, etc.) or a comparison with other risks.

Examples of comparisons with environmental standards include occupational exposure limits (OEL), acceptable daily intake (ADI) values for various substances, or technical reference concentration values. In these cases, emissions and exposures are compared with standards so as to decide on risk management measures. The number of currently enforceable environmental standards in Germany is very large and these standards are extremely heterogeneous in both their rationales and their aims.[2] This is not to disparage the practice of such comparisons but to indicate that such a method of evaluation lacks transparency and can also be inconsistent.

Besides the comparisons with environmental standards, there are comparative evaluations of different risks with the goal of creating a risk ranking. This type of evaluation, termed comparative risk assessment (CRA), has been both discussed and applied in practice in the USA since the end of the 1980s. The most well known and perhaps most ambitious CRA project is the ranking of environmental risks carried out by the US Environmental Protection Agency (US EPA). In addition, numerous CRA projects have been, and are being, carried out in the USA, mostly on behalf of particular states or at the local level. The goal of these efforts was, and is, to lay the foundation for a more rational establishment of priorities in environmental health policy. Using scientific risk assessments, risks from entirely differ-

---

[2] In its analysis of environmental standards the German Advisory Council on the Environment (SRU) recorded 154 lists of environmental standards with a total of approximately 10,000 individual environmental standards; see SRU (1996).

ent areas are to be evaluated and compared so as to arrive at an estimation of their relative significance.

In contrast to the USA, comparative risk evaluations have as yet been only rarely used in Germany (and Europe), either in the field of risk assessment and evaluation or in risk management.[3] There are multiple reasons for this: for one the methodological difficulties inherent in CRA are cited, while another is the completely diverging estimation of the benefit of CRA for environmental policy.

This neglect of CRA is perhaps also a reflection of the density of regulations and the philosophy of regulation followed in Germany, in which risks from various regulatory areas (nonionizing radiation, ionizing radiation and chemicals, transport, etc.) and in various fields of environmental policy (protection of wildlife, soil, water, and climate, as well as air purity conservation, waste disposal, and the treatment of hazardous materials) are not considered concomitantly. This sector-by-sector consideration has disadvantages, not the least of which is the insecurity perceived by citizens when a risk, which is not tolerated in one field, is accepted in another. This public insecurity also results from the basic principle followed in Germany of regulating technical facilities through safety certifications without producing any quantitative reference to risks. In particular, the isolated approaches of individual environmental health sectors regardless of a sufficient consideration for the overall context of safety and risk need to be changed. This is where CRA, when properly integrated into the process of risk regulation, can perform a useful service.

The claim that comparative risk evaluation forms a rational foundation for making environmental health-related decisions has led, especially in the USA, to a critical debate in which a range of arguments have been brought against CRA. One set of arguments relates to the use of CRA in the context of the USA's regulation of environmental health risks (in particular on the part of the US EPA). It is, for instance, argued that the use of CRA would be accompanied by a "tyranny of the rational", which would reduce the flexibility of the existing system of regulation in the USA and that this flexibility is precisely the result from the fact that risk assessment and risk management are not separated from the political processes (Silbergeld 1995, p. 422).

Other criticisms are of a more fundamental nature. In addition to the accusation of the trivialization of risks – especially of environmental risks which, by being compared to so-called "lifestyle risks" (e.g., smoking, or nutritional habits) are often accorded a minor fraction of the risks facing human health – three arguments in particular are brought against CRA (e.g., Hornstein 1992; Shrader-Frechette 1995; Silbergeld 1995).

---

3) Morgenstern et al. (2000) in their international comparison of CRA methods and results mention a series of developing countries, but find no use of CRA with the aim of developing risk ranking in Europe or, indeed, the OECD countries. CRAs on specific themes have, in the meantime, been attempted in single research projects, for instance the GaBE Project on the Risks of Energy Supply Systems (Projekt zu Risiken von Energieversorgungssystemen); cf. Hirschberg et al. (1998). See Section 3.2.5 of this book.

The first argument asserts that risks are frequently so qualitatively heterogeneous, for instance in respect to their genesis or the type of damage they cause, that a meaningful comparison cannot be made. This is the famous "you can't compare apples with oranges" problem. However, the fact remains that apples and oranges can indeed be compared quite well, for example in regards to their flavors, vitamin contents, or prices. The only requirements are, firstly, that the objects to be compared share at least one property in common and, secondly, that this property is seen as meaningful for the purposes of comparison. In this respect, the argument about incommensurability draws attention to an important aspect of the comparative assessment of risk: to be meaningful, any comparison of risks must take into account the aspects of the risk, which are relevant to the evaluators. One can, therefore, readily compare the risk of smoking with that of driving if one accepts the statistical expectancy of annual deaths through smoking and driving as the relevant characteristic for comparison.

The second argument centers on the considerable knowledge gaps about environmental health risks and deduces that comparative evaluations of these risks are not possible or, at the least, not meaningful. Indeed, risk assessments often bear considerable uncertainties, stemming both from the limitations of theoretical understanding and from a shortage of empirical data. This is a serious problem, not only for a CRA but also for any form of evaluation and regulation of risks. Nonetheless, decisions about how to deal with risks must be, and are, made. Any form of evaluation and regulation of risks must appropriately consider the uncertainties; how this can occur within the framework of a CRA is one of the central themes of this book.

Finally, the third argument sees in CRA an obstacle to the possible participation of the public in decisions on environmental health risks. Relying on comparative risk assessment for such decisions is seen to create the danger of an "expertocracy", because the risk assessments upon which a CRA must be based are frequently very complex and, thus, can hardly be followed by nonexperts. Even the implementation of a comparative risk evaluation demands methodological know-how. However, it is precisely the process of CRA that offers the opportunity to connect scientific expertise on the one hand with the values of societal groups, i.e., stakeholders, on the other in a transparent and systematic manner. Experts have the task of supplying information about the nature, magnitude, and likelihood of risks (together with the associated uncertainties), while the evaluation of this information, as well as the selection and weighting of assessment criteria, can be performed by the societal groups or stakeholders.

Streffer et al. (2000, p. 15) point out that three kinds of knowledge are required in order to be able to compare risks in a rational way. Firstly, one must be able to assess the nature and probability of consequences. In other words, the causal, or at least conditional, relationships must be known. Secondly, one must know which of the consequences the societal actors who are involved in the comparison of the risk (e.g., experts, political representatives, those affected by the risk) regard as harmful. In the case of obvious harm (death or illness) these are beyond dispute, but other damaging effects (e.g., well-being disturbances, or those damages expected to occur in the distant future) may give rise to differing opinions among the actors as to their detrimental nature. Thirdly, knowledge of how risk comparisons are to be conducted is essential. The scientific/technical process of determining risks, i.e., risk assessment, is not meant but rather the methodology of comparative risk evaluation and its assumptions.

It is this third kind of knowledge that concerns the following final report of the BfS project Risk Evaluation and Management an Elaboration of Plans for an Integrated and Comparative Approach to Risk. The project's goal was to develop a methodological foundation of the processes of integrated comparative risk evaluation, which could do justice to the various demands that may arise within the context of risk regulation. We have therefore focused above all on the issue of the comparative evaluation of risk with the objective of risk ranking. This book is not about the comparative evaluation of action alternatives for risk management with the goal of their prioritization – for which additional aspects need to be considered, such as the cost efficiency of alternative measures. Comparative risk evaluation, however, constitutes the foundation of these further evaluations.

The use of the term "integrated" signifies that risk, and thus also comparative risk evaluation, is a multidimensional concept, which comprises various aspects that must be explicitly dealt with. Risk can refer to various hazards which can have different effects (for health, the environment, the economy, etc.) and for whose evaluation differing criteria may be brought to bear. Risk assessments are performed by different experts or groups of experts who may reach divergent assessments of a risk. These estimations could then be evaluated differently by different stakeholders.

This starting position means it is essential for the design of a CRA to take into account not only the scientific foundation of risk assessment but also the issue of how a CRA process is to be implemented. Thus both methodological challenges and questions regarding the process management arise. Insights from the fields of behavioral studies, social sciences, and decision-making can be drawn upon to provide the scientific basis for the procedural competence required for process management.

Objectives

The objectives of this study are to evaluate existing approaches to CRA, to explore the possibilities of improving such comparisons, to discuss the opportunities for participation, and to design processes that will increase acceptance of political decision-making about environmental health risks.

The following questions are addressed:
- What prior efforts of comparative risk evaluations are there? What gaps and problems do these demonstrate?
- What types of comparative risk evaluation can be distinguished, and which of these, within existing administrative constraints, are meaningful and worthy of further development?
- On what formal structures can a comparative risk evaluation be based?

- How are the individual steps of a comparative risk evaluation to be designed, and what advice can be gleaned from the substantial body of literature on social science risk research?
- What special demands does an integrated comparative risk evaluation have, in particular one that involves a variety of evaluators from science and society?
- What are the chances of success for comparative risk evaluations, and where do their limitations lie?

These goals have given rise to the structure of this book:
- Chapter 2 clarifies certain core definitions and concepts (risk, hazard, risk assessment, etc.)
- Chapter 3 presents sample cases from the spectrum of CRA processes, which have been conducted so far.
- Chapter 4 gives an overview of the results of social science research relating to risk and decision-making which can be used in the organization and implementation of CRA processes and their requisite communication of risk.
- In Chapter 5 the methodological problems associated with CRA are examined and the approach of multiattribute decision-making as a method for use in comparative risk evaluations involving a variety of evaluative criteria and involved parties is presented.
- Chapter 6 discusses the practical organization and implementation of a CRA.
- The appendices contain more detailed information on risk metrics and explanatory examples of multiattribute comparative risk evaluation, as well as on the comparative evaluation of unclear risks and risks arising from hazardous incidents.

2
Concepts and Definitions

In this chapter the major concepts used in the discussion and elaboration of a comparative approach to risk are explained. Additional definitions and comments are found in the appendices and glossary at the end of this book.

2.1
Risk

The terms chance and risk both entail the idea that a specified possibility can become reality. This is mostly connected with the idea that the change from a possibility to a reality can in some way be evaluated. Whether such a possibility will actually be realized, however, remains unknown. The unknown is a constituent property of chances and risks. When we can only be imprecisely aware of what is possible, then an additional uncertainty results. Given a specific set of circumstances, this uncertainty can by itself result in a risk.

Risks usually give rise to negative expectations. In general terms, a risk means an occurrence with undesirable effects or, expressed in another way, the possibility of a harmful occurrence. The specific conceptualization of this general idea is, however, different within various scientific disciplines.

Jungermann and Slovic (1993, p. 169f) identify six differing definitions of risk, of which definitions (d) to (f) are primarily used within the fields of financial and actuarial mathematics:
(a) risk as the probability of a harmful event;
(b) risk as the magnitude of a possible harmful event;
(c) risk as a function (usually a product) of the probability and magnitude of a harmful event;
(d) risk as the variance of the probability distribution of all the possible consequences of a decision;
(e) risk as the semivariance of the distribution of all negative consequences with a specified point of reference;
(f) risk as a weighted linear combination of the variance and the expected value of the distribution of all possible consequences.

Comparative Risk Assessment. Holger Schütz, Peter M. Wiedemann,
Wilfried Hennings, Johannes Mertens, and Martin Clauberg
Copyright © 2006 WILEY-VCH Verlag GmbH & Co. KGaA, Weinheim
ISBN 3-527-31667-1

Technical risk analyses, or systems analyses, are usually conducted for facilities that could represent a substantial hazard to their environment, mostly because of the use of "materials which, through their toxic or radioactive effect or through energy release can cause damage to health or to property" (Hauptmanns et al. 1987). Such facilities possess safety systems, which should enclose such materials or prevent such energy releases. Technical risk analyses, or systems analyses, thus place emphasis on the possible and initially unknown courses of hazardous incidents and the aspect of system failure. "The concept of risk is used to assess and evaluate uncertainties associated with an event. Risk can be defined as the potential for loss as a result of a system failure" (Ayyub and Bender 1999, p. 7).

Even within the sphere of health, each scientific field uses its own variants of the terms associated with risk. For the area of cancer research, for instance, Williams and Paustenbach (2002, p. 368ff) state that "risk is a unitless probability of an individual developing cancer". Within epidemiology, risk is customarily expressed as relative risk, odds ratio, and attributable risk. Risk is here the relation of observed cases of harm in one population to another population. In toxicology, risk is understood as "the likelihood, or probability, that the toxic properties of a chemical will be produced in populations of individuals under their actual conditions of exposure" (Rodricks 1992, p. 48). In similar vein, the WHO and the United Nations Conference on the Human Environment define risk as "the expected frequency of an undesirable effect caused by exposure to a contaminant" (adapted from Neubert 1994, p. 848).

Kaplan and Garrick have supplied a definition of risk which is independent of any particular field and which highlights three significant elements of risk. They have defined risk as a set of triplets (Kaplan and Garrick 1981, p. 13):

$$R = \{\langle s_i, p_i, x_i \rangle\}, \quad i = 1, 2, \ldots, N \tag{2.1}$$

where s_i is a scenario identification or description; p_i is the probability of that scenario; x_i is the consequence or evaluation measure of that scenario, i.e., the measure of damage.

Risk is thus described within this conceptualization as the product of the consequences with their possibility and magnitude.[4] The European Commission's Scientific Steering Committee's Working Group on Harmonization of Risk Assessment Procedures takes up a similar conceptualization of risk:

> The probability and severity of an adverse effect/event occurring to man or the environment following exposure, under defined conditions, to a risk source(s). (EC 2000b, p. 5)

This is the definition of risk that is used in this book.

Several authors have suggested an expansion of the scientific and technical concept of risk with so-called "qualitative" risk factors (Fischhoff et al. 1984; Florig et

---

[4] Kaplan and Garrick (1981) and Kaplan (1997) differentiate this conceptualization further. However, this definition suffices for the purposes of the following discussion.

al. 2001).[5] These "qualitative" risk factors are based on the results of research into risk perception (see below). These have shown that two aspects are significant in the perception of risk: the extent to which the source of a risk (technology, product, material, etc.) is perceived as something to be feared, and the degree to which the source of risk is seen as an unknown for science or the public. Fischhoff et al. (1984), for instance, distinguish three dimensions of risk: mortality (in which they further distinguish between workplace-related mortality and that in the general public), morbidity, and concern. This last is an umbrella term for the qualitative dimensions of risk, namely "dread" and the "unknown" (see Figure 1).

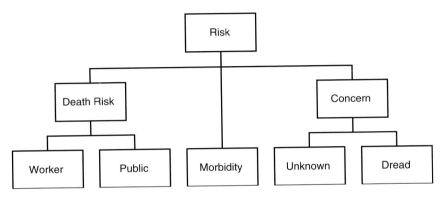

Figure 1
Multidimensional concept of risk of Fischhoff et al. (1984, p.129).

Another approach places the application of concepts of risk within the legal sphere.[6] Risk here is considered as falling within a conceptual triad: danger, risk, and residual risk. Exact legal definitions of the terms risk and danger, it must be said, are as yet unavailable. Danger is spoken of as "when the circumstance of an unchecked course of events may be expected to lead to events with a sufficient probability of harm, i.e., to non-negligible damage to legally protected goods and property" (SRU 1999, p. 39). Dangers must, under German law, be averted.

Risk, from another point of view, is understood as any possibility where the certainty of harm or damage is not sufficient to confirm the existence of a danger. Finally, the term "residual risks" denotes those risks that are to be tolerated by society. These involve, as the German Federal Constitutional Court's famous Kalkar verdict fundamentally formulated, uncertainties beyond the threshold of common sense, which have their origins in the limitation of human cognition.

It follows from this that there are, in principle, also unknown risks and unknown dangers. In the areas of both risks and dangers, it is possible to imagine cases of

---

5) The WBGU (1998) has also, through its classification of risks into various types, carried out such an expansion, in which, among others, ubiquity and potential societal mobilization are applied as classificatory criteria.

6) These comments on the legal concept of risk are largely based on the SRU (1999).

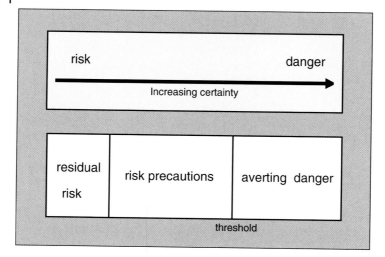

Figure 2
Risk, danger, and the associated concepts of environmental law.

which we as yet know nothing, but which cannot logically be excluded from the realm of the possible. Figure 2 shows the relationship between risk, danger, and the associated concepts of environmental law.

2.2
Danger/Potential for Damage–Hazard

Contrary to the usage in colloquial speech, where danger and risk are frequently employed as synonyms, the two concepts are differentiated in the context of scientific risk assessment. Danger here means the potential for a source of risk (technology, product, substance, situation, activity, etc.) to cause adverse effects.[7] Sociological risk research also distinguishes between risk and danger, particularly with respect to the materialization of a negative expectation (Beck 1986). A danger is something threatening or strange that occurs, for example a natural catastrophe. A risk, in contrast, is a possible future consequence attendant upon the choice of a particular course of action, and is, therefore, mainly considered to be open to influence.

In this connection, then, it is more appropriate to speak of "potential for damage" or "hazard" in order to define this potential more closely and thus arrive at an assessment of risk. Danger in this context is also to be distinguished from the con-

---

[7] The WHO (1994, p. 20) gives the following definition of the term adverse effect: "change in morphology, physiology, growth, development or life span of an organism which results in impairment of functional capacity or impairment of capacity to compensate for additional stress or increase in susceptibility to the harmful effects of other environmental influences. Decisions on whether or not any effect is adverse require expert judgment."

cept of danger as a quantitatively defined factor, as used in the field of law (see above).

2.3
Incertitude and Uncertainty

Incertitude and uncertainty are intrinsic to the concept of risk. These terms are often used as synonyms or are not properly differentiated from each other. In the context of comparative risk evaluation, however, it is useful to distinguish these terms clearly.

Incertitude refers to a circumstance in which we cannot know whether something we have identified as a possibility will become reality. Incertitudes of this sort can also occur when the possibilities of their occurrence as such are well known; examples of this include the individual throws of dice. In contrast, the term uncertainty is applied as a measure of how precisely the possibilities of these occurrences – including, where applicable, their statistically expected frequencies – are known.

On occasion, these two cases of uncertainty are distinguished as aleatory (in the sense of an incertitude, as above) or epistemic (caused by a lack of knowledge). When drawing conclusions from comparative risk evaluations, it is of substantial significance whether an individual risk is of aleatory or epistemic "origin".

2.4
Risk Assessment

Risk assessment is understood as the "process from identifying a potential danger to the quantitative characterization of risks" (German Risk Commission 2002, p. 9). Different procedures will be employed depending upon whether the assessment is of environmental or health risks or of risks associated with a technical system.

Within the assessment of health risks, four steps are most usually distinguished (cf. National Research Council 1983). The first step is the identification of a hazard, i.e., the qualitative identification of the harmful properties of a material. This concerns, for instance, the question of whether a substance can cause cancer. The second step is the quantitative description of the dose–effect relationship. Which effect is produced by which dose is ascertained and, above all, whether there is a threshold value for harmful effects or whether there are risks associated with even the smallest doses. The third step constitutes the assessment of exposure, i.e., the strength and duration of exposure to a harmful substance that a population faces. Finally, in the fourth step a characterization of the risk is undertaken. In other words, a summary evaluation is given in which the type and frequency of the expected harm to the health of the exposed population is presented.

The assessment of ecological risks involves the assessment of the possible adverse effects on an ecosystem of exposure to one or more stressors. An ecological risk assessment may be divided into three phases (US EPA 1998a). In the first phase, *problem formulation*, the appropriate section of the ecosystem is defined and the ecological effects and the stressors assumed to be the causes are described in a (usually qualitative) conceptual model. This forms the basis for the second phase, *analysis*, in which the probability of exposure is tested and the magnitudes and types of ecological effects expected from a particular level of exposure are determined. From the resultant exposure and stressor effect profiles, a summary *risk characterization* is then performed. This risk characterization includes the associated assumptions and the assessment uncertainties inherent in the analysis.

When determining the risks associated with technical facilities, one needs to differentiate between a plant's normal operation and the possible occurrence of hazardous incidents. In the case of normal operation where the emission of risk-relevant material is the issue to be reckoned with, the corresponding risks may be evaluated as above. An analysis of the risks associated with breakdowns requires specific additional procedures. Initially, this involves a statistical compilation and modeling of the individual components of a technical system (cf. Mertens 1993). Methods such as fault tree or event tree analyses are to be applied here. In this way the failure performance of the whole system can be calculated with the corresponding probabilities. Finally, the risks resulting from individual failure situations (such as health risks or the ecological risks of emissions stemming from a hazardous incident), which can likewise be characterized by probability distributions, are specified.

2.5
Risk Evaluation

Risk evaluation is built on the results obtained from scientific risk assessment. While risk assessment has the aim of determining the size or magnitude of a risk, risk evaluation assesses the risk in terms of its significance, gravity, or seriousness. The size or magnitude of a risk gives no indication of how "bad" the risk is. For instance, a relative risk of two (i.e., a doubling of the risk for an exposed person relative to that of an unexposed individual) of getting a particular disease tells us nothing about how significant that might be. This significance is only revealed when the risk is compared either to a reference value (such as an environmental standard) or to another risk. A risk evaluation always implies, in other words, a comparison. This raises the question as to who should conduct the risk evaluation. For cases of individual risk-taking, that is, for risks that only affect the person who takes the risk (a situation which occurs most often with so-called "lifestyle" risks), the answer is simple: the person affected should conduct the evaluation.

The issue is more problematic when it is that of the societal regulation of risks. Here the question is about the tolerability or acceptability of risks.[8] The past has

---

[8] What is under discussion here is the acceptability, rather than the acceptance, of risks. While the latter is a term that describes the empirically ascertainable approval of risks on the part of people, groups, or society, the former means the postulation that the risk in question can be approved. Acceptability is, therefore, a normative concept.

shown that a universally applicable solution for this cannot be found. How great a risk may be – or how small a risk must be – to be societally acceptable so as to be tolerable to its citizens is an issue on which there is as yet no societal consensus.[9] Before one can address, within the framework of societal risk evaluation, how the acceptability of a risk can be justified, two other problems must be solved.

Firstly, the dimensions considered in a scientific assessment of risk (damage and probability) are not the only aspects that can be considered as meaningful to the evaluation of risks by various societal actors. As a matter of fact, in the past, risk evaluations made use of the same dimensions that were applied in scientific risk assessments. These are the risk metrics used in various disciplines: cases of death, specific endpoints (diseases), and toxicological or epidemiological parameters.[10] Psychological and social-scientific research into risk perception has, however, shown that the intuitive appraisal of risk by lay people can involve a whole series of so-called "qualitative" factors. These are related to the nature ("quality") of the harm or damage and to the situational context in which the risk occurs (Jungermann and Slovic 1993). As above in the elaboration of the concept of risk, it is here suggested that these qualitative aspects should be made explicit within the conceptualization of risk and, consequently, also in risk assessments. In its report *Welt im Wandel – Strategien zur Bewältigung globaler Umweltrisiken* (Changing World: Coming to Terms with Global Environmental Risk), the WBGU takes a similar approach. It suggests "consulting scientific assessments together with the risk perceptions uncovered by empirical studies as the basis of a rational consideration (Fiorino 1989). Both types of information are important components of risk evaluation" (WBGU 1998, p. 40). In contrast to the WGBU, we do not consider such an inclusion of risk perceptions within risk evaluation to be sensible; in our opinion, these should be based exclusively on scientific risk assessments. However, during the prioritization of risk management measures, particular groups' or the public's risk perception can, of course, become a relevant criterion of judgment.

Secondly, risk assessments seldom yield unambiguous results. As a rule, they suffer from more or less considerable uncertainties.[11] The problem thus arises as to how these uncertainties may be incorporated into risk evaluation. In the first place, these uncertainties have to be acknowledged within the risk evaluation, a point that is by no means self-evident. The results of risk assessment are often expressed as point estimates without any characterization of the concomitant uncertainties being reported. This practice has been much criticized in recent years (e.g., National Research Council 1994). It is argued that estimations expressed as single point estimates tend to be based on worst cases and thus overestimate risks. This can lead to the misallocation of resources for health or environmental care (Finkel 1989). Uncertainties can exist in all stages of a risk evaluation. It is possible, for

---

[9] Cf. in this connection the extensive discussion in WBGU (1998, Ch C 1).

[10] For toxicology these are, for example, unit risk, benchmark dose, margin of exposure; and for epidemiology relative risk, odds ratio, attributable risk.

[11] For a condensed discussion of this problem, which has been frequently revisited in the literature of risk assessments, see, e.g., Bailar and Bailer (1999) and US EPA (1996).

instance, in the analysis of dose–effect relationships and exposures to generate uncertainties within quantitative data and then express these as confidence intervals, percentages, or probability distributions (cf. Thompson and Bloom 2000). In contrast, uncertainties in the identification of potential damage may only have a limited quantitative description. The core problem here is that these uncertainties arise from limited or contradictory sets of findings. One can try to collect the findings together quantitatively with the aid of meta-analyses. In any case, this places great demands on the quality and comparability of available data.[12]

These two problems, the multidimensionality of risk evaluation and the approach to uncertainty, are central to risk evaluation and will be referred to repeatedly within this work.

2.6
Risk Comparison

A risk comparison has the aim of correlating two or more risks on the basis of risk assessments. Clear objectives must be formulated to arrive at useful results. Such objectives usually arise from problems with related questions.

One such problem arises, for instance, when comparatively classifying a risk with the goal – for example in relation to specific health endpoints – of conveying its significance. Here one is concerned primarily with the understanding of, and communication about, risk. Another objective for a risk comparison can be to put risks on a scale so as to permit a well-founded setting of priorities for decisions about risk management in circumstances of limited resources, or to arrive at a cost–benefit viewpoint. Finally, a risk comparison can help to appraise different options (such as the choice between several technical solutions or courses of action) with regard to their risks. Here risk is one of a number of factors that may be weighted or aggregated as necessary.

From a systematic point of view one can position comparative risk evaluation between risk management and risk assessment. Figure 3 presents this relationship in the form of a pyramid. Risk management forms the apex of this pyramid. Questions at this level about the relative significance of risks determine the context of the comparative risk assessment. In the performance of this, the necessary information about the risks under consideration is extracted from existing risk assessments and any further necessary risk assessments are initiated. On the basis of these data the comparative risk evaluation is conducted and its results provided as a risk ranking. The methodology of a comparative risk evaluation, as well as questions regarding its organization and conduct, are dealt with in detail in Chapters 5 and 6.

This ranking in turn forms the basis of a prioritization of risks. The prioritization of risks is, thus, to be distinguished from the risk ranking. The prioritization

---

[12] See, for the problem of the summary characterization of groups of findings in the identification of potential damage, Wiedemann et al. (2002, Ch 7).

Figure 3
The CRA pyramid.

requires further reflection, for instance whether measures are even possible under the prevailing conditions or which criteria should be used to decide on risk management measures (cost efficiency, fairness, etc.). Comparative risk evaluation is, therefore, just one – albeit a fundamental – factor in the prioritization. The choice of criteria for the appraisal of risk management measures and the appraisal itself are discrete issues, which may well require the involvement of various stakeholders and interest groups.

2.7
Risk Management

Risk management concerns itself with those activities that are related to dealing with risks. It occurs after risk evaluation, which clarifies how significant a risk is. The decision concerning whether and how a risk will be regulated or controlled is then the task of risk management, because following the risk evaluation other factors are still relevant. Risk management and risk evaluation are, therefore, separate and independent concepts, which fulfill different functions within the process of risk regulation (see below).

In the first place, the decision whether risk management measures are even necessary must be made. The choice of protection strategy is of fundamental importance: whether the danger will be explicitly averted (protection from danger) or precautionary measures taken (precaution). As discussed above, if damage is to be prevented by averting the danger, one must reckon with sufficient certainty on its appearance without such measures. Choosing precaution requires as a minimal prerequisite that in the process of risk evaluation a scientifically unambiguous determination of potential damage can be made. Criteria and measures are specified in laws and regulations for the protection of health and environment and the safety of technical systems. In contrast, precaution must tackle the question of whether and how to deal with great uncertainties within scientific risk evaluation. These uncertainties are chiefly associated with the identification of potential damage, such as the question whether a particular material has a harmful effect on health. The protection strategy of precaution is also the subject of some laws (e.g., the federal laws governing emissions, chemicals, and genetic technology).[13] The significance of this for risk management lies primarily in societal or political arguments between interest groups, where one seeks to promote the precautionary principle and the other to reject it.

Following the decision that measures should be taken in response to a risk, courses of action need to be developed. Health, ecological, economic, technical, social, and political points of view need to be taken into consideration. The next step is the evaluation of the courses of action according to criteria related to, for instance, the efficiency of these options in minimizing risk, the cost–benefit aspect, and the societal enforceability of the measures. A further important aspect within the evaluation is the possibility of undesirable consequences ensuing from these courses of action. This information forms the basis for the decision about the risk management strategy to be implemented. Risk management also involves overseeing the implementation of measures and subsequently evaluating their effectiveness.

Fundamentally, the ways of dealing with risk can be divided into four strategies: avoidance, reduction, aftercare, and protective coverage (Wiedemann 1993, p. 209).

Avoidance means the removal of all possibility of harm during the use of technologies, products, substances, etc. This is normally achieved at societal level through laws or regulations. Examples include the banning of chemicals or the setting of limits (e.g., of exposure). Such limits only avoid risks where a limit value exists in the dose–effect relationship of the toxin in question. This is the case for most noncarcinogenic materials. In contrast carcinogens exhibit a (mostly linear) dose–effect relationship without a threshold value. With such pollutants risk cannot, in principle, be avoided by setting limits, only reduced.

The strategy of *reduction* aims not to remove risk but to minimize it. This can happen through a lessening of the incidence rate of harmful events. In the case of carcinogens, this means reducing exposure as far as possible;[14] in the case of technical facilities it can be achieved by applying appropriate safety precautions. Otherwise, measures to limit the extent of damage can be taken, for instance by limiting exposure in the workplace.

[13] Cf. Rehbinder (1988) and Wiedemann et al. (2001, Ch 4).

[14] The ALARA (as low as reasonably achievable) principle was originally developed to address this problem.

Aftercare signifies measures that help to cope with the consequences of a harmful event. These include a broad spectrum of heterogeneous measures, from the preparation of infrastructure necessary in the case of harm to emergency financial support and psychosocial treatment of affected individuals. In the main these are strategies for extreme events such as natural or technological catastrophes and have as yet received little attention within the field of environment-related health risk management, although in recent years such attention has been increasing.[15]

The strategy of *protective coverage* has an entirely different aim. This is about the compensation of affected individuals for harm, loss, or damage. This is usually done by taking out insurance, and sometimes also through the issue of state guarantees/bonds. One example of the latter is the state's coverage of loss and damage due to accidents involving nuclear power, which exceeds the sum for which the power plant operator is insured, or (for a certain period following the attack on the World Trade Center on 9 September 2001) against the risk of terrorist attacks on civil aviation.

Above all, the problem of risk–risk tradeoffs arises from the strategies of avoidance and reduction (Graham and Wiener 1997). This refers to the fact that the avoidance or reduction of a risk can itself have risks among its consequences. An illustration of this is the cleanup of old waste sites, which aims to remove or reduce the risk of cancer to residents from the carcinogenic materials at the site. However, such cleanups bear the risk of accidents to workers who perform the cleanups, which can be weighed against the reduction of health risk to residents (Cohen et al. 1997).[16] The usual analytical instruments for making decisions about risk management options, cost–benefit analyses, and cost-effectiveness analyses are focused on the cost–benefit relationship or the cost effectiveness of risk management options. However, whether other risks may accidentally arise from risk management measures is mostly left out of consideration by such an approach. Such risk displacements are presumably not isolated incidents; they should be examined using risk–risk tradeoff analyses during the appraisal and selection of risk management options (Graham and Wiener 1997; Presidential Commission 1997, p. 35).

2.8
Risk Regulation

When dealing with risks in society, risk assessment, risk evaluation, and risk management in many ways occur together. However, at the same time they constitute a sequence of necessary steps for the rational regulation of risks. The German Risk Commission has depicted the entire process from risk assessment through risk evaluation to risk management as risk regulation (see Figure 4).

As a first step, the German Risk Commission sets a preliminary procedure at the beginning. A series of decisions are to be made here regarding the delimitation of

---

15) Cf. Kunreuther (2002).
16) Graham and Wiener (1997) term these neglected risks, which can work against risk reduction through risk management measures, hence "countervailing risk".

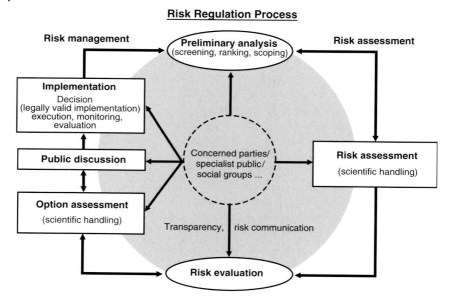

Figure 4
The process of risk regulation (German Risk Commission 2003, p. 23).

the problem and the controlling and timing of the phases of the process to follow. These include organizing the early identification of risks, the determination of the basic conditions for the later risk assessment, a setting of priorities – perhaps in especially urgent cases – as well as the possible choice of a shortened regulation procedure, for instance for resource efficiency, if greater risks can be excluded from the start.

The second stage of the process consists of the risk assessment. The purpose of this scientific analysis is the identification of the potential hazards of a pollutant, the quantitative evaluation of the hazard, an estimation of exposure, and a statement of the associated uncertainties.

In the next step, the German Risk Commission calls for a risk evaluation. The results of the risk evaluation are discussed and evaluated in terms of their consequences for risk management. The scientifically specified risk is "translated" into societal terms (German Risk Commission 2003).

In the fourth stage, risk management, a possible call for action is determined, possible measures for risk reduction are identified and appraised, and the impacts of these measures – such as in the effects on different subjects of protection – differentiated in a comparative evaluation. Finally a recommendation is made to administrative and political decision-makers.

In the opinion of the German Risk Commission, this whole process should involve not only experts but also other groups from society:

> In a society in which pluralism of values prevails and where there
> is always great pressure to justify political actions, risk
> evaluations often meet with scepticism or distrust. For this
> reason, statements about risks are more dependent than other
> statements on plausibility and on trust in the regulating bodies.
> The more individuals and groups have the opportunity to play an
> active part in risk regulation, the greater is the chance that they
> will come to trust the institutions and also assume responsibility
> themselves.
>
> [...] This being so, timely reciprocal participation of the parties
> concerned and the organised social groups in the decision-finding
> process is not only objectively appropriate, but also called for from
> a constitutional and democratic point of view.
>
> (German Risk Commission 2003, pp. 47–48)

2.9
Risk Communication

Communication about risk plays a central role in a transparent account of the assumptions, uncertainties, considerations, and findings of the individual steps in the process of risk regulation. This is true not only of communication between those directly involved in the risk regulation process (experts and societal groups) but also for the general public. Such communication, risk communication, is thus far more than the simple exchange of information between involved parties. The US National Research Council states:

> Risk communication is an interactive process of exchange of
> information and opinion among individuals, groups, and
> institutions. It involves multiple messages about the nature of
> risk or other messages, not strictly about risk, that express
> concerns, opinions, or reactions to risk messages or to legal and
> institutional arrangements for risk management.
>
> (National Research Council 1989, p. 21)

As a first approximation, two goals of risk communication can be distinguished: enabling exchange and ensuring comprehension. As a result two different groups of tasks can be identified. *Exchange* relates to the processes of enabling and maintaining lines of communication. At issue here are two-way communication, participation, as well as ensuring fairness, respect, and trust. *Comprehension* means assisting the understanding of the technical fundamentals of risk evaluation on the part of the recipients and thus contributing to their risk literacy.

So as to prevent dilution of the meaning of the term "risk communication", further use of this expression will be limited to circumstances in which significant cognitive, motivational, or emotional differences between information provider and recipient prevail. This is particularly the case when experts communicate with lay people, but such differences can also be assumed to occur between experts when these come from widely differing professional backgrounds.

3
CRA in Practice

Approaches to integrated and comparative risk evaluation can only be properly appraised when one's perception of their utility includes and considers the varieties of CRA currently existing in practice. This chapter therefore deals firstly with the various evaluations of and positions on CRA in the USA, Europe, and Germany. This is complemented by four examples of comparative risk evaluation that are presented and discussed.

3.1
Political Environment of CRA

3.1.1
CRA in the USA

Making comparative risk evaluations a foundation of environmental, and in a certain sense also health, policy was an innovation of the USA's Reagan/Bush administrations in the 1980s. The topic has remained on the political agenda in the USA since that time.

Not only Congress and the federal authorities but also state and local policy were deeply concerned with the problem of allocating resources for the protection of environmental health. This was originally related to the deregulation of environmental policy through a more rational setting of priorities ("the biggest bang for the buck"). CRA was seen as an instrument for steering strategy, i.e., as a decisive control quantity between risk evaluation and risk management.[17]

The institutions established in the 1970s, most notably the Environmental Protection Agency (EPA), the Occupational Safety and Health Administration (OSHA), and the Consumer Product Safety Commission (CPSC), were criticized for spending large parts of their budgets addressing marginal risks that were, in comparison with other risks, hardly significant (cf. Breyer 1993). They were censured

---

[17] "Comparative risk analysis is intended principally as a policy-development and broad resource-allocation tool" (US EPA Science Advisory Board 2000, p. 5).

for being so strongly guided by political considerations and public risk perceptions that the objective risk situation (by this was meant the risk assessments of experts) was largely neglected in the shaping of environmental policy for the allocation of funding resources.[18]

The first comparative risk evaluation project initiated by the US EPA, Unfinished Business, aimed "to develop a ranking of the relative risks associated with major environmental problems that could be used as one of several important bases on which EPA could set priorities" (US EPA 1987).

Following the Unfinished Business project a whole range of similar projects were implemented, the underlying problems of such comparisons much discussed (cf. Finkel and Golding 1994), and a variety of suggestions for their improvement made (Konisky 1999), e.g., within the framework of the EPA's Integrated Risk Project. The report Toward Integrated Environmental Decision-Making of the US EPA Science Advisory Board (2000) suggested one particular procedure. According to the characteristics revealed by the answers to these three questions: (a) is the risk low, high, or uncertain?, (b) is the potential for a risk reduction high?, and (c) is risk reduction economically viable, equitable and supported by the public?, one of the following measures should be selected:
- monitoring of any change in the risk
- research or education
- risk management monitored for its effectiveness.

The appropriate involvement of all societal groups is explicitly called for.

> *Although integrated decision-making requires the involvement of a broad spectrum of participants (e.g., scientists, engineers, economists, decision-makers, and the public), the different groups have unique roles to play. In other words, the Framework does not imply that "everyone must be involved in everything all the time." For example, just as scientists are not expected to provide the perspective of the general public, members of the general public are not expected to conduct the technical analyses of scientists.* (US EPA Science Advisory Board 2000, p. 11)

Approximately 100 CRAs have been performed to date. Projects are being implemented in practically every state.[19]

The results of the first CRA projects in the USA also indicated some shortcomings, even though the procedure had been followed. The goal of dividing scarce resources according to the significance of risks was not reached. From the current perspective, however, one can observe that the political debate about CRA settled

---

[18] The experiments of Hood et al. (2001) from the Centre of Risk Regulation of the London School of Economics showed similar tendencies in England to those of the USA. The rigor and extent of risk regulation were largely determined by public risk perceptions.

[19] The Environmental Defense Fund offers the best overview. CRA projects are also to be found in Canada, New Zealand, Hungary, Poland, the Czech Republic, and Southeast Asia.

down and that new facets emerged, particularly through the discussion of the precautionary principle.[20] It would appear that positions in the USA with regard to the applicability of comparative risk evaluation have polarized.

Between 1993 and 1998 the US Congress (House of Representatives and Senate) dealt with over a dozen legislative initiatives concerning risk evaluation. There is no doubt that policy efforts to create a comprehensive legal framework for risk evaluation and CRA were advantageous for the further development of CRA. However, only the most minor of these initiatives were adopted by both chambers.

Progress in the application and resultant political legitimization of CRA was achieved through executive orders. The first initiative originated from the Reagan administration in 1981, which through Executive Order (EO) 12291 (Regulatory Planning Process, 50 Federal Register 1036) transferred the supervision of cost–benefit analyses of statutory regulations to the Office for Management and Budget. Under the Clinton administration, EO 12291 and EO 12498 were replaced by EO 12866 and EO 12875 (Regulatory Planning and Review, 58 Federal Register 51735, 1993), in which CRA was explicitly named, but appropriate procedures were not definitively prescribed. EO 12866 said that

> ...each agency shall consider, to the extent reasonable, the degree and nature of the risks posed by various substances or activities within its jurisdiction...

Furthermore, the performance of cost–benefit evaluations was viewed as one of the obligations of the authorities. They were to

> ...[assess] all costs and benefits and select those approaches that maximize net benefits, incorporating economic and distributive impacts and equity considerations into the analysis. Agencies should adhere to the principle that the regulations should be pursued in the most cost effective manner, the benefits should justify the costs, and the regulation should be based on the best scientific, technical, economic and other information...

Consequently, it can be concluded that CRA has established itself as one of the instruments of priority setting in the USA.

3.1.2
CRA in Europe

CRA in Europe has nothing like the prominence that it enjoys in the USA. In the EU Commission's first report (EC 2000a,b) on the harmonization of methods of risk evaluation the methodological difficulties of CRA were observed and the exis-

---

[20] The chief political criticisms were for being overly technocratic in approach and preventing, or at least hindering, public participation in risk-related decisions. Other causes for concern were that CRA could limit the flexibility of environmental action, that it over-prioritized lifestyle risks and so pushed environmental risks off the agenda (cf. Commoner 1992; Hornstein 1992; McCloskey 1994; see also Minard 1993).

tence of greatly different estimations of the utility of CRA in forming environmental policy were noted. Within the report itself, this utility was given a cautiously positive appraisal and a further engagement with the topic was called for.

Nonetheless, there are also approaches to comparative risk evaluation in Europe. In some spheres, such as the setting of priorities in the chemicals field, risk-ranking procedures have been suggested (cf. Hansen et al. 1999). The risk due to human exposure to a chemical is expressed by the exposure (in the first instance by the tonnage, the manner of use, and the physical properties of the chemical released) and by the toxicity and is documented in R-Phrases.[21] Something similar is to be found in the EU Commission's Chemicals White Paper (EC, 2001). Here a new system of assessment, REACH (registration, evaluation, authorization of chemicals), is suggested: depending on the tonnage and properties of the material, a registration, evaluation, and authorization process is conducted. A priority assessment is provided for chemicals available in quantities of more than 100 tons per year. The same goes for substances that are deemed to be dangerous on the grounds of their inherent properties (such as substances that can alter the genome, very toxic chemicals, substances with highly persistent or bioaccumulative characteristics, and substances whose molecular structures give cause for concern). For particularly dangerous materials, those that give "cause for grave concern", an authorization procedure should be performed. Those substances obliged to be authorized include persistent organic pollutants (POPs) and substances that possess demonstrably carcinogenic, mutagenic, or reproductive toxic properties.

One pan-European project comes from the European Environmental Agency (EEA), which in 1995 published the Dobris Assessment about the state of the European environment.[22] Some 56 various environmental problems were evaluated. The EEA used the following attributes: (a) threat to sustainability; (b) impact of global problems in the European area; (c) prominence of a European aspect; (d) transboundary aspect; (e) long-term character; (f) risk for human health; (g) social or cultural impacts; (h) ecological damage; and (i) economic loss.

The report does not explain how these attributes were applied and weighted. A priority list was presented which contains no gradations. The EEA thus identifies the following 12 environmental problems as particularly pressing: climate change, damage to the ozone layer, loss of biodiversity, technological and natural risks, acidification, low-level ozone, inland water, damage to forests, marine, and coastal environments, waste materials, the urban environment, and chemical substances. A second status report was published in 2001 in which progress regarding these environmental problems was assessed.

---

[21] The EURAM rankings are intended to provide a reference and a focus for the discussions leading to the preparation of priority lists. They should be seen as a convenient tool for summarizing the vast amount of information in IUCLID in such a fashion that an expert can use this data summary to identify those substances that are expected to exhibit the greatest risk to people or the environment. See http://ecb.jrc.it/existing-chemicals/

[22] English: http://reports.eea.eu.int/92-827-5122-8/en/ (German: http://reports.eea.eu.int/92-827-5122-8/de/).

The EEA does not clearly enough distinguish risks, in the sense of possible events, from existing damage. A glance at the list of environmental problems, furthermore, makes clear that the problems not only differ in the range of their concepts but also occupy different positions within the causal chain of a risk event, from cause to final effect. Thus causes of risks (e.g., biotechnology or the transport and storage of hazardous wastes) are to be found on the list alongside impacts (e.g., soil degradation or damage to the ozone layer).[23]

Various European authorities have recently discussed CRA. So, for instance, the European Medicines Evaluation Agency in 2001 relied on a "comparative risk/benefit reassessment" to revoke partially the authorization of the drug Orlaam.[24]

In Great Britain the Advisory Committee on Pesticides has concerned itself with the issue of comparative risk evaluation and addressed the topic in its meetings of 2001.[25]

The Swedish government's Ministry of Education and Science has referred critically to the lack of debate about comparative risk evaluation in its statement to the EU's 5th social framework program:

> *Furthermore, there is an important need for environmentally oriented socio-economic and behavioural research. A comprehensive perspective, comparative risk assessment and risk management as well as developing legal instruments, financial mechanisms and incentives are important aspects of environmental research...*[26]

The Standing Committee on the Food Chain and Animal Health of the European Commission (General Health and Consumer Protection – Directorate E) declared itself in favor of the use of comparative risk evaluation in October 2002.[27]

---

23) Nuclear accidents; stratospheric ozone; depletion; increasing UV; loss of biodiversity and genetic resources; resources and quality of groundwater; acidification; hazardous waste (transport and storage); climate change; forest degradation; waste disposal; nuclear waste; urban air quality; nature conservation and sensitive ecosystems; persistent air toxics; industrial accidents; direct pollution (discharges, dumping) to sea; soil contamination from waste disposal; conservation of threatened species; tropospheric ozone increase and episodes; waste production; soil and resource contamination; fragmentation and destruction of habitats; eutrophication of surface waters; riverine inputs into seas; changes in hydrological regime; management of large rivers and lakes; desertification; lack of water supply; stress and degradation due to tourism; food security; urban waste; growing vulnerability of complex systems; bioaccumulation (metals, POPs); energy security; soil erosion; risks of biotechnology; microbiological pollution of surface water; oil spills; sea-level rise; intensification of land-use; introduction of new organisms; shortage of industrial raw materials; coastal erosion; natural radioactivity (radon); loss of agricultural land; shift of biogeographical zones; draining wetlands; floods; droughts and storms; nonionizing radiation; societal health; landscape modification; noise; occupational health; loss of cultural heritage; seismic activity volcanoes; pests and locusts; thermal pollution of waters.

24) http://www.emea.eu.int/pdfs/human/press/pus/877601en.pdf
25) http://www.pesticides.gov.uk/acp.asp?id=520
26) http://www.cordis.lu/fifth/src/ms-s-1.htm
27) http://europa.eu.int/comm/food/fs/rc/scfcah/plants/rap12_en.pdf

The European Food Safety Authority, discussing research activities into acrylamide, also called for comparative risk evaluation of "health risks from heat treated foods and food products".[28]

In the European Commission (General Health and Consumer Protection – Directorate C)'s latest report (2003), the Scientific Steering Committee Task Force approved comparative risk evaluation.[29]

Overall there are indications of a trend towards a positive evaluation of CRA.

3.1.3
CRA in Germany

CRA is as yet uncommon in Germany. One of the exceptions is the report of the LAI's working group (LAI 1992). The group worked on a study lasting several years, Evaluative Criteria for Defining Cancer Risk due to Air Pollution, which was published in 1992. The study dealt with the following substances: arsenic and arsenic compounds, asbestos, benzene, cadmium and cadmium compounds, diesel exhaust particulates, PAH (benzo[a]pyrene), and 2,3,7,8-tetrachlorodibenzo-p-dioxin (TCDD). The assessment is based on "unit risk" values, which indicate the estimated additional cancer risk to an individual of constant exposure over 70 years to a concentration of 1 µg of toxic substance per cubic meter of air.

However, the idea of CRA in the form of a programmatic comparison of risks – as it is carried out in the USA – meets with considerable resistance.

A taxonomy of risk has been suggested as an alternative. In 1998, for instance, the WGBU presented an approach to risk classification that distinguished six different types of risk. The types are differentiated by these attributes: (a) probability, (b) magnitude of harm or damage, (c) certainty of probability analysis, (d) certainty of analysis of harm or damage, (e) persistence, (f) ubiquity, (g) delay in action of effect, (h) irreversibility, and (i) potential mobilization of the public. However, a prioritization of the various types was not attempted; rather, a range of risk management strategies were derived from the risk classification.

One interesting aspect of the WBGU process is that the procedure it suggests permits the assessment of the most heterogeneous sources of risk. To this extent the process resembles programmatic risk comparisons. In contrast, however, the risk sources are not ranked, but assigned to various treatment categories that contain no gradations concerning the setting of priorities. The WBGU process is not, though, without difficulties in its application, because actual risks do not readily fall into the WGBU's typology as it may seem at first sight. In addition, the management strategies for the various risk types do not remain so different when their operationalization is attempted. How, for instance, can a distinction be made between precaution and long-term responsibility that are identified as two distinct options?

A second approach, which contains at least fragmentary comparisons, was presented by the German Advisory Council on the Environment (SRU 1999). Accord-

---

[28] http://europa.eu.int/comm/food/fs/sfp/fcr/acrylamide/study_area10.pdf
[29] http://europa.eu.int/comm/food/fs/sc/ssc/out325_en.pdf

ing to this, the impairments to health caused by allergies, ultraviolet radiation, or public noise had been somewhat underestimated in policy. In contrast, the SRU asserted that an overestimation of the impairments to health due to substances producing hormonal effects and disease patterns caused by multiple chemical hypersensitivity had occurred in the public debate. These appraisals were offered by experts (the members of the SRU), but no explicit comparative methodology is to be found in the report.

The Action Programme for the Environment and Health (APUG 1999) of the Federal Ministry of Health and the Federal Ministry for the Environment, Nature Conservation and Nuclear Safety also discussed the most important environmental problems (noise, air, radiation, water, soils and contaminated sites, settlement hygiene, foodstuffs and articles of daily use, products, hazardous incidents, and selected environmental toxins). However, the APUG authors did not determine a risk ranking.

The German Risk Commission requests that risks be assessed systematically in the future. The multitude of competencies and working levels in the various institutions means

> *that the way potential health risks are dealt with is usually determined by a fortuitous or interest-driven perception of the problem on the part of the public or the media. [...] Hesitant and contradictory regulation by the governmental bodies involved, selective information of the parties concerned, and communication deficits combine to produce a situation where on the one hand relatively minor risks occupy an important position in public perception, whereas on the other, serious risks are underestimated or even "swept under the carpet".*
> (German Risk Commission 2003, p. 20)

The German Risk Commission sees the functional role for risk comparisons firstly as risk evaluation and secondly as risk communication. They express their view thus:

> *Comparisons of risks in public risk communication are of questionable value. However, for a rational prioritization of the application of public and private resources for minimizing or preventing risks, they are essential. Scientific risk assessment can merely provide quantitative estimates of incidence rates and (quantitative) estimates of the trustworthiness of the assessment. Even the question of the comparability of different adverse effects, i.e., the quality of risks, lies beyond the realm of the empirical-cognitive and is mostly determined through the perception of values.* (German Risk Commission 2002, p. 58 [translated from German])

Despite all the skepticism about CRA, comparing and prioritizing risks seems almost unavoidable. Nonetheless, no single methodologically elaborated project exists in Germany as yet, which compares risks through various media and endpoints as is the case in the USA.

This lack is in our opinion due on the one hand to the density of regulation and to the prevailing philosophy of regulation on the other, which does not consider risks in various regulatory areas (nonionizing radiation, ionizing radiation, and chemicals, transport, etc.) and in various fields of environmental policy (protection of wildlife, soil, water, and climate, as well as air purity conservation, waste disposal, and the treatment of hazardous materials) together. In addition, the evaluation of safety, rather than of risks, is at the forefront of German regulatory practice. A CRA would only make sense in Germany if it were implemented from the viewpoint of precaution and risk minimization.

3.2
Project Case Studies

3.2.1
US EPA Unfinished Business Project

Aim of the project
Comparative risk evaluation arose in 1987 from the EPA's requirement to prioritize risks from various regulatory spheres for the purposes of allocating resources. This was accompanied by an examination of the appropriateness of the EPA's former expenditures for various environmental programs.

Procedure
In 1986 EPA administrator Lee M. Thomas appointed an internal task force of 75 US EPA scientists and managers to investigate the "relative risks to human health and the environment posed by various environmental problems". The task force's assignments were defined and divided as follows:

1. The entire sphere of environmental problems (including issues not necessarily under EPA's jurisdiction) was classified into 31 problem areas. The division followed the existing programs' affiliations and regulations. So, for instance, the topic blocks "...air pollutants, hazardous air pollutants, contaminants in drinking water, abandoned hazardous waste (e.g., Superfund) sites, pesticide residues on food, and worker exposures to toxic chemicals..." were selected from the existing programs.[30]
2. Four areas of risk were distinguished: cancer risk, noncancer risk, ecological effects, and social ("welfare") effects. It is to be noted that in Unfinished Business social effects do not concern the whole societal environment but primari-

---
[30] For this reason the term "programmatic comparative risk assessment" is used to describe a comparative risk evaluation which considers the risks of one or more programs.

ly material, as, for instance, "...damages to property, goods and services or activities to which a monetary value can often be assigned. These include natural resources (e.g., crops, forests and fisheries), recreation (e.g., tourism, boating), materials damage and soiling (e.g., building materials), aesthetic values (e.g., visibility) and various public and commercial activities..." The risk areas were evaluated by four different working groups and their findings were not collated.
3. The following aspects were not addressed by the project: (a) the economic or technical controllability of a risk, (b) the qualitative aspects of risks which are often regarded as particularly important by the public, such as the degree of voluntariness involved in taking the risk, the knowledge about the risk or social equity issues, and (c) the social utility of the activity associated with the risk.

Further steps in Unfinished Business tackled the problem of integrating quantitative data and qualitative information ("expert judgment") as well as the establishment of joint analytical guidelines for the four working groups. Within the groups, metrics were selected which for cancer risk (number of cases) and social effects (financial loss) were pre-existing; for noncancer risk and ecological effects, however, new metrics had to be developed.

A "summary sheet" was structured for each of the 31 problem areas by each of the four working groups, in which existing information on risks, as well as sources and uncertainties, were to be recorded. The summary sheets were filled in by the appropriate program units and evaluated within the working groups. Although the working groups "successfully developed a spirit of striving for objectivity in this project", subjective assessments – even on the part of experts – could hardly be avoided.

These limitations of the comparative risk project Unfinished Business were not considered to detract from the quality of the results to the extent that they called into question the result of the ranking:

> The method we used to compare environmental problem areas can best be described as systematically generating informed judgments among Agency experts. About 75 career managers and experts representing all EPA programs participated over a period of about nine months. The participants assembled and analyzed masses of existing data on pollutants, exposures, and effects but ultimately had to fill substantial gaps in available data by using their collective judgment. In this sense the project represents expert opinion rather than objective and quantitative analysis. But despite the difficulties caused by lack of data and lack of accepted risk evaluation methods in some areas, the participants feel relatively confident in their final relative rankings.

The procedure of the Unfinished Business project is summarized in Figure 5.

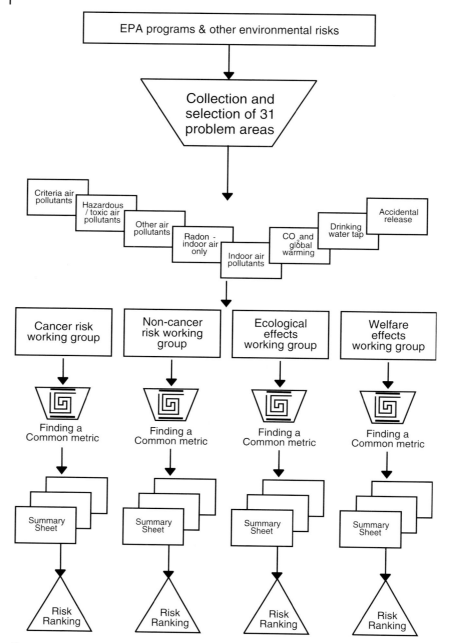

Figure 5
Procedure of the Unfinished Business project.

Selection of different subjects of protection, influences, and effects to be considered
Conventions played a role in the development of the 31 problem areas and in the selection of chief attributes for the CRA in the four working groups. According to Unfinished Business, there could as easily have been 11 or 111 problem areas. The project team decided on a number of about 30 to provide a balance between the pursuit of detail and the need for manageable analysis. How the risks were categorized is indeed no minor matter, because the larger the problem area, i.e., the more individual effects counted therein, the further up the ranking the problem area was likely to be placed.

In addition, the collation of the problem areas can lead to an overlap of effects, so that a "double counting" of risks was possible (see also Morgan et al. 2000).

> *Some of these problems are diffuse mixtures of sources effects and a variety of different problems. This has led to some "double counting" of risks in more than one problem area. For example health risks from inactive hazardous waste sites (problem area #17) often result when people either drink contaminated water or inhale volatile toxic chemicals. Such health risks will be double counted by the drinking water problem area (#15) or by the hazardous air pollutant problem area (#2)...*

One property of environment-relevant pollutants can be that they spread rapidly through a variety of environmental media, a characteristic that can also lead to double counting:

> *...when we were able to, we tried to take account of intermedia transfers and secondary effects. Some pollutants easily cross the boundaries of environmental media (e.g., air, water). Sulfur dioxide (SO_2) emissions ... covers risks from SO_2 in the air ... as well as damages from the deposit of sulfates on structures ... and from their eventual arrival in water ... Essentially, we attempted to follow the pollutant "from the cradle to the grave." Secondary effects occur when pollutants such as SO_2 are chemically transformed ... and do damage in a new form.*

The report emphasizes that the selection of environmental problems could also have been made on the basis of consistent criteria, such as sources, contaminants, paths, or receptors. Moreover, it was mentioned that the "intangible" facets of an effect, such as degree of voluntariness or social equity, could not, due to a lack of concepts, data, and metrics, be integrated into the analysis.

These limitations belong to those problems that must be endured. Others rest on a lack of data sets, which can only be remedied by expert opinions, which themselves are known to bring problems.

Dose–effect relationships
A consistent approach to dose–effect relationships and to exposures proved to be difficult to achieve. This point was expressed within the report:

> EPA analyses have made differing assumptions about emissions of various pollutants, the exposure that people, ecosystems and objects receive and dose/response relationships. Comparing estimates generated under incompatible assumptions was difficult. In some instances where there were sufficient data, we tried to coordinate assumptions. In other instances, we simply tried to judge the magnitude of the bias produced by the incompatible assumptions and compared risks without reworking the underlying quantitative estimates.

In some cases a consideration of the different dose–effect relationships was attempted. However, cross-group coordination was lacking, so that the risks calculated by the working groups for cancer risks and noncancer risks, ecological, and social effects were not based on a unified set of assumptions about dose–effect relationships.

Selection of affected regions and population groups
The conclusion of the report draws attention to the fact that this national study does not necessarily reflect local circumstances and that local comparative risk evaluations are needed:

> This analysis is not a guide to what may be the most serious problems in a particular area or for particular individuals. Any attempts to set local priorities should take into account local conditions.

Furthermore, the differences in the vulnerability of various population groups or ecosystems were not included.

Selection of timeframes considered
In Unfinished Business the EPA set the timeframe for the period of the study:

> To facilitate a fair representation of the damages from all environmental hazards, we tried to estimate all the damages that will occur from the problem as it exists now.

Future developments in hazards were, thus, not brought into consideration, and neither were the differences in duration of the latency periods before the onset of adverse effects:

> ...many of the effects associated with various air pollutants take place at essentially the same time as they are emitted into the environment ground water contamination from hazardous waste sites may not affect human health for years, even decades, after contaminants have leached from a site ... Instead of discounting or using some other method of indicating time preference, we simply presented damages in terms of their magnitude, sometimes with a notation about the time at which they occur.

Valuation of damage

While the metric for cancer risk (number of cases) and the metric for social effect (financial loss) could be based on established methods, new common metrics of comparative risk evaluation had to be developed for ecological effect and noncancer risk.

The so-called "intangible" effects, such as "options values" and "existence values", which relate to the utility of a resource, were only partially brought into consideration:

> *In many instances, people value environmental quality to a greater degree than the tangible benefits they derive from a clean environment would seem to suggest. This is because of the intangible characteristics of the risks. People place intrinsic value on a clean environment. The public's vehement insistence on cleaning up contaminated ground-water aquifers even when they are not currently being used is an example of these important intangible values.*
>
> *The fact that we did not fully consider these intangible characteristics in this project should in no way suggest that they are unimportant. In fact a number of studies have found that they are a significant component of what the public perceives as the total value derived from environmental protection. Rather we did not fully consider them here merely because not enough work has been done in this area to enable the project participants to assess them fairly for all the environmental problems examined in the project.*

Stakeholders

Only scientists and managers from the ranks of the EPA took part in the Unfinished Business project. Other stakeholders, in particular the general public, were not included.

Conclusion

The US EPA's Unfinished Business project earned the distinction of being the first project to focus on addressing the issues of comparative risk evaluation. The deficiencies visible from the point of view of today were also to some extent seen by the project's participants and acknowledged as inherent problems to be borne.

Of particular significance is the division of risks into problem areas oriented on the contemporary EPA structures. In combination with the selected spheres of risk this led to some double counting of risk. Here it is significant that, along the entire causal chain of risk events, from cause through exposure to effect, factual, circumstantial, or even latent cross-connections could occur which could have a decisive impact on the design of comparative evaluations and the appraisal of their findings.

An analysis of the projects highlights the importance of a unified set of assumptions regarding dose–effect relationships, the inclusion of benefit in an examination of risk, the problem of consistent evaluation of diverse sorts of harm or damage, and is in particular critical in respect to the question whether a purely financial assessment of social aspects is sufficient.

Within the procedure the following points are highlighted:
- The risks (problem areas) were divided into four spheres (cancer risk, risk of other diseases, ecological effects, and welfare (societal) effects); these were subsequently dealt with separately.
- In the case of cancer risk and material welfare effects, standard recognized metrics were already available. With their assistance the risks could be quantitatively assessed.
- In the case of noncancer risk and ecological effects, metrics had first to be developed. On account of inadequate data, these risks were only qualitatively included in either three or six categories.

In the overall evaluation, none of the four spheres were viewed as more important than the others, and no attempt was made to produce an aggregate ranking. In the separate deliberations of the four spheres, four clusters were identified (US EPA 1987; Morgenstern and Sessions 1988):
- a high rank in three of the four spheres, or at least a middle ranking in all four spheres;
- a high rank in health risks, but low ranking in ecological and welfare risks;
- a high rank in ecological and welfare risks, but low ranking in health risks;
- only middle or low ranking throughout.

Of these, the first three clusters were viewed as important, the last less so.

Table 1 gives an overview of the procedure.

3.2.2
Comparative Risk Framework Methodology and Case Study (US EPA)

Background
The report (US EPA 1998b) was expressly designed as a preliminary study and has not yielded any fully developed methodology.[31] This is also expressed in the review of the report (US EPA Science Advisory Board 1999).

Aim of the study
The EPA's comparative risk evaluation methodology was conducted on the basis of a hypothetical case study. The case study deals with a comparison of various methods of disinfecting drinking water. The background to the case is the realization that the previously prevalent chlorination of water forms organic chlorine compounds, which can cause health risks. The risks of disease due to bacterial infec-

---

[31] "INTERNAL DRAFT: DO NOT CITE OR QUOTE – This document is a preliminary draft. It has not been formally released by the US Environmental Protection Agency and should not at this stage be construed to represent Agency policy. It is being circulated for comments on its technical accuracy and policy implications."

Table 1 Procedure of the US EPA's Unfinished Business project.

Objective	Priorities for the allocation of resources (personnel and funding): public health
Spheres of risk	Cancer risk Risk of other diseases Ecological effects Material social effects
Comparison categories	Problem areas according to the internal organizational structure of the EPA
Timeframes	Future harm (e.g., cancers that may only manifest after a latency period) were treated as though they had arisen immediately, i.e., they were not discounted
Geographic frame	Regional peculiarities were expressly not considered
Attributes	Cases of death, diseases, ecological effects, welfare effects
Aggregation/metric	Four separate spheres of risk, categorized within each sphere: Cancer risk: 26 quantitative categories Risk of other diseases: 3 categories Ecological effects: 6 categories social welfare effects: 24 quantitative categories The risks were evaluated by experts forming informed judgments, i.e., considering all the facts including uncertainties. In many cases, lacking data and methods, the rank of the risk was subjectively estimated
Overall evaluation	Four clusters: a high rank in three of the four spheres, or at least a middle ranking in all four spheres a high rank in health risks, but low rank in ecological and material risks a high rank in ecological and welfare risks, but low ranking in health risks only middle or low ranking throughout
Uncertainties	Uncertainties were considered in principle, but were frequently not quantified due to a lack of data. They were assessed qualitatively
Presentation of results	Lists of rankings

tion following ineffective disinfection and the risks of disease due to disinfection byproducts (DBP) for established and alternative disinfection procedures were considered and compared.

Model
The fundamental model calculates on the one hand the germs which, depending on the quality of the untreated water, remain in the drinking water and, on the other hand, the chemical compounds resulting from disinfection procedures. The health effects were calculated and converted into a common health metric, the Quality Adjusted Life Year (QALY). Figure 6 shows a schematic of the model.

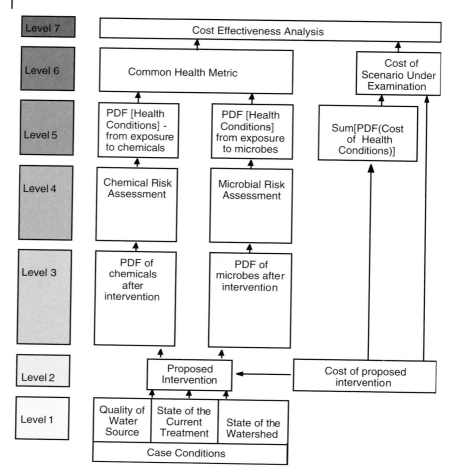

Figure 6
Schematic overview of the Comparative Risk Framework Methodology model. PDF = probability distribution function.
(After US EPA Science Advisory Board (1999).)

Choice of technology
Following the objectives, the case study deals with an assessment of methods and standards of disinfection of drinking water. Accordingly, the standard disinfection process at the time was compared with two modified procedures. The influence of water distribution systems was neglected (this was criticized).

Selection of subjects of protection, influences, and effects to be considered
The selection reflected the known harmful effects of germs and pollutants. The affected subjects of protection dealt exclusively with human health.

The case study considered:
- Only the path through direct consumption of drinking water (and not, for example, through skin or by inhalation while washing, showering, and bathing).
- Only the direct consequences for the consumer of drinking water; the infection of third parties by those already infected was addressed but not (here) considered. This limitation was criticized by the US EPA Science Advisory Board (1999).
- Only the effects of normal services, but not the effects of disruptions of services; this limitation was criticized by the US EPA Science Advisory Board (1999) because disruptions, i.e., errors in disinfection facilities, had on several occasions in the past led to infection epidemics.

The following limitations arose from the conceptual formulation:
- No consideration of effects beyond those for directly affected individuals, in particular no ecological effects were considered.
- Only consideration of the direct effects of disinfection procedures, not of the effects arising from, foe example, the construction of disinfection facilities or the manufacture of disinfection agents.

Selection of affected regions and population groups
The calculations were performed for a specific location. Two groups were examined, one a normal population and the other a particularly sensitive group of people with weakened immune responses.

This was criticized by the US EPA Science Advisory Board (1999): to be able to establish a standard, a representative set of locations and representative population groups must be used. The population groups in the case study considered to be particularly sensitive are too small a proportion of the population as a whole, so that, when only the particular sensitivity of this group is considered, the effectiveness of water preparation may be underestimated. This criticism corresponds to the following characteristic of the procedure. For each of the alternatives investigated in the comparison, the QALYs of the population as a whole were ascertained by summing the QALYs of those not particularly sensitive with the QALYs of the particularly sensitive individual subpopulation. Supposing, however, that the proportion of particularly sensitive individuals within the population as a whole is small, say 1%, then an alternative which reduced this group's QALYs by, e.g., 20%, while the ordinary population were not affected, would lead to a decrease of $1\% \times 20\% = 0.2\%$.

Dose–effect relationships
The calculations in the case study were performed using known (or considered as proven) dose–effect relationships. These were accompanied by substantial uncertainties, for instance regarding the applicability of the results of animal experiments to humans. For more details, see the section entitled "Uncertainties" below.

Valuation of damages

To make a comparison of different methods possible, various types of damage must be converted to a unified comparative scale. This is a necessary fundamental phase of assessment that demands particular transparency, because it inevitably includes subjective evaluations.

In *Comparative Risk Framework Methodology and Case Study* this is the QALY. This approach is based on the idea it makes a difference whether it is a healthy individual or an individual weakened by sickness that dies as a result of a detrimental effect. The approach attempts to consider this by giving every stage of life a weighting factor between 0 and 1, in which 1 signifies "perfect health" and 0 "death or complete loss of quality of life due to illness". This allows disease and death to be converted to a common scale.

This procedure raises the following problems:
- Who defines the life quality factor of a given illness? What is perfect health? (Does it even exist?)
- Are there not also diseases that are perceived to be "worse than death"? Must these be given a negative weighting factor?
- The second question indicates that there is a question of one's WTP (willingness to pay) implicit in the QALY concept: "How many years of life would you be willing to sacrifice to be spared this illness?" It is unclear here whether years of life from the same period of life are meant (which would involve a corresponding moving forward of the aging process) or a bringing forward of the time of death in an otherwise identical lifetime. Thus the evaluator has to compare a situation that may be concretely envisioned with a somewhat theoretical concept (loss of vaguely specified life years).
- The WTP question leads to a discounting question: "would it make a difference at what time of life an illness struck?"
- How should acute symptoms be evaluated? Due to their brevity, these may have been more narrowly evaluated than the subjective estimations of affected individuals (US EPA Science Advisory Board 1999; Hubbel 2002, p. 7).
- When weighting the possible years of life lost due to acute burdens on health, it must be remembered that the particularly sensitive individuals may be so sensitive precisely because they have previously suffered a long-term chronic burden. If one reduces the years of life lost resulting from the acute burden due to the individuals' poor state of health without taking into account that this poor state of health in turn was a result of the chronic burden caused by the same source, then not only is the chronic burden ignored but the acute burden is assigned an incorrectly low weighting. This is not a failure of the QALY approach, but it shows the necessity of dealing with diseases and cases of death together and not as separate outcomes (Hubbel 2002, pp. 9–10). "Does air pollution just cause death in already ill people, or does it cause the disease that leads to death?" (Hubbel 2002, p.21).

The QALY metric imposes several restrictive assumptions (Hubbel 2002, p. 5):

> QALYs are consistent with von Neumann–Morgenstern utility theory only if one imposes several restrictive assumptions, including independence between longevity and quality of life in the utility function, risk neutrality with respect to years of life, and constant proportionality in tradeoffs between quality and quantity of life.

Uncertainties

The study primarily considered the uncertainties of numerical parameters. (Their values can be established relatively easily and their influence on the results can be ascertained with relative ease.) The US EPA Science Advisory Board (1999, p. 29) criticized the absence of any reflection on uncertainties within the model. The report should have contained at least a qualitative discussion of the possible magnitude of uncertainties within the model. Furthermore, it was recommended that a basic distinction be made between variability (in the form of distribution, e.g., of a population) and actual uncertainties, and that this distinction be presented alongside the results.

One example of uncertainty within a model which plays a role not only here but generally within risk analyses is the extrapolation of dose–effect relationships from animal experiments involving high doses to low doses among humans. This is associated with great uncertainties.

There is a need for further research concerning methods for calculating the probability distribution of states of health (level 5 in Figure 6) from calculated risks (level 4 in Figure 6) (US EPA Science Advisory Board 1999, p. 11).

Conclusion

Many of the problems that arise from comparative risk evaluations do not appear in this case study due to its conceptual formulation. Alternatives that produce qualitatively homogeneous benefits and involve risks, which are qualitatively homogeneous, should be compared. Thus the bacterial risks associated with two alternatives may be compared directly and without problems; this is also the case with chemical risks. Problems arise only when one alternative is associated with a high chemical risk and the other with a high bacterial risk. Then risks of different natures must be compared, which may also be of different magnitudes for different population groups. This is where the common metric plays a decisive role.

The US EPA Science Advisory Board (1999) judged the Comparative Risk Framework Methodology to be a "potentially powerful tool that is well suited to helping identify important variables that are presented by complex environmental problems" (p. 10). The QALY concept met with fundamental approval, but should have been compared with other metrics to select the most appropriate and defensible weightings developed for the chosen metric. The choice of metric should be clearly justified.[32]

---

[32] "Even though the Committee is not questioning whether the QALY is the most suitable metric, it believes that alternatives should be considered, adequately discussed in the document, and a clear case made for the metric selected" (US EPA Science Advisory Board 1999, p. 10).

These stipulations result from the fact that a risk comparison is based on the metric and on evaluations which cannot be entirely scientifically derived but must be based on other approaches.

Finally, it is essential that the uncertainty of results must be considered in decision-making. The possibility cannot be excluded that in the comparison of two risk values, the uncertainty of one risk or of both is greater than the difference between these values. When this is the case, the values cannot form the basis for a decision.

Table 2 gives an overview of the procedure.

3.2.3
Global Burden of Disease (GBD) Study

Aim of the project
The Global Burden of Disease (GBD) project was initiated by the WHO working in conjunction with the Harvard School of Public Health and the World Bank in 1992 (GBD 1996). The aim of the project was a comprehensive analysis of the burden of disease and injury on the world population. The endpoints examined were prema-

Table 2 Procedure of US EPA Comparative Risk Framework Methodology and Case Study.

Objective	Comparison of alternative technical methods; "public health" and protection of the individual
Spheres of risk	Health risks due to disinfection of drinking water
Comparison categories	Various technical methods (in the form of scenarios)
Timeframes	Regarding emissions, the length of life of the facility concerned, here set at 20 years; regarding health effects, the lifetime of the exposed individual
Geographic frame	Supply area of the facility
Attributes	Cases of death, illness
Aggregation/metric	Various illnesses and cases of death were aggregated using the QALY metric
Overall evaluation	Implicitly multiattribute due to the use of the QALY metric
Uncertainties	Variability and uncertainty both considered for the numerical parameters in the calculations. Uncertainties within the model not considered.
	Resultant uncertainties of results calculated with Monte Carlo simulation. However, uncertainties only given for detailed results, not in summary presentation of results
Presentation of results	Various detailed results, itemized for alternatives to be compared various illnesses various population groups and with variants for different discounting Summary result as a comparison of aggregated QALY values for the alternatives

ture deaths and disabilities; each differentiated by age group and gender and divided into eight geographical regions of the globe. Among many characteristics, the death rates for more than 100 diseases and injuries, nearly 500 manifestations (processes and consequences), and the influence of the 10 most important risk factors were ascertained. Scenarios projecting the development of premature deaths and disabilities were elaborated. In a follow-up project, GBD2000, the methods by which the original project's insights were gained were further developed and made useable for national purposes (GBD 2000).

The goal of the project was to create a reliable foundation from which health authorities could make use of data sets of varying quality and completeness for a global evaluation of states of health and the causes of disease and disability and for future health policy. Lopez and Murray (1998) describe this goal as:

> ...to weave these disparate data sets of variable reliability and completeness into a global assessment of health conditions and the causes of disease and injury to guide public policy into the next century.

From this main goal, three specific objectives were derived for the GBD project:
1. A systematic incorporation of more information than the mere number of deaths into the assessment of states of health. To this end, a time-based measure of the number of years of healthy life lost, whether caused by premature death or understood as years of life with a disability, either as a result of disease or injury, should be applied. Thus, the extent of disability is adjusted using a weighting factor.
2. All estimates and extrapolations should be based on the most objective epidemiological and demographic methods possible.
3. The measurement of the burden of diseases or injuries should be achieved through the use of a metric that also enables the estimation of the cost effectiveness of measures or interventions.[33] The metric chosen is disability-adjusted life years (DALY).

Procedure
A novel quantitative indicator, which represented the burden both of premature death and nonlethal patterns of disease, was developed for the project. This indicator is based on the principle of years of life lost, adjusted for quality of life as well as the adverse effects and disability resulting from disease or accidents. The indicator, described as DALY, is the sum of years of life lost through premature death (YLL) and years of life lived with disability (YLD), based on societal circumstances at any one time:

$$DALY_i = YLL_i + YLD_i$$

in which i represents the current disease situation.

[33] The project's success in achieving this objective resulted in the widespread adoption and application of the DALY indicator, such that this "objective" indicator rapidly gained the status of a normative measurement.

A DALY means a year of healthy life lost. The prematurity of a death is calculated from the current difference between statistical lifespan for the society with the highest life expectancy (Japan).[34] DALY values were calculated for 108 different diseases or sources of injury causing death or disability according to ICD-9 (9th Revision of the International Classification of Diseases). These diseases and sources of injury (hereinafter referred to as GBD causes) were grouped into three main categories, infectious or parasitic diseases, noncommunicable diseases, and injuries, which were in turn grouped into 9 subcategories[35] and broken down into a total of 483 categories of sequelae.

This procedure is illustrated in Figure 7. The left-hand section of the diagram contains a schematic representation of the procedure as a series of single steps. The right-hand part shows the quantitative form given to each of the GBD project's steps.

Murray and Lopez (1996a) enumerate four aims that they sought to achieve through the development of the DALY indicator:

1. The development of internally consistent estimates for mortality due to 107 primary causes of death,[36, 37] differentiated by age group and gender and divided into eight geographical regions of the globe.
2. The development of internally consistent estimates for the incidence, predominance, duration, and specific mortality of 483 disabling sequelae resulting from the causes mentioned above, differentiated by age group and gender and divided into geographical regions.
3. The evaluation of the proportion of mortality and disability ascribable to the 10 most significant risk factors, again differentiated by age group and gender and divided into geographical regions.
4. The development of scenarios for mortality and disability, divided according to cause and, once more, by age group and gender and geographical region.

The risk factors are categorized along the aspects of nutrition, consumption and miscellaneous habits, environmental influences, working conditions, and similar. Examples include vitamin deficiency, smoking and alcohol, air pollution, noise, and also physical inactivity (Ezzati et al. 2002).

An important point to note is that the indicators are dependent on societal values. Weightings are necessary, for example for the question how grave a certain disability is, or whether a year of life as a young adult is worth more or less than a year

---

[34] In the opening phases of the GBD study Japanese life expectancy was used as a factually occurring maximal value. However, both a theoretical life expectancy for humans and, in particular, a practical life expectancy for a certain population can be derived from dynamic statistics of life and death. This was partially used in GBD2000 and for the national Burden of Disease projects.

[35] The number of clusters for causes of death was raised to 21 subcategories for GBD2000. http://www3.who.int/whosis/menu.cfm?path=whosis,burden,burden_manual&language=english, p. 17.

[36] GBD causes are meant here.

[37] The number of diseases and injuries investigated as causes of death was given as both 107 and 108 in the original literature, and is usually given as 108 in later references.

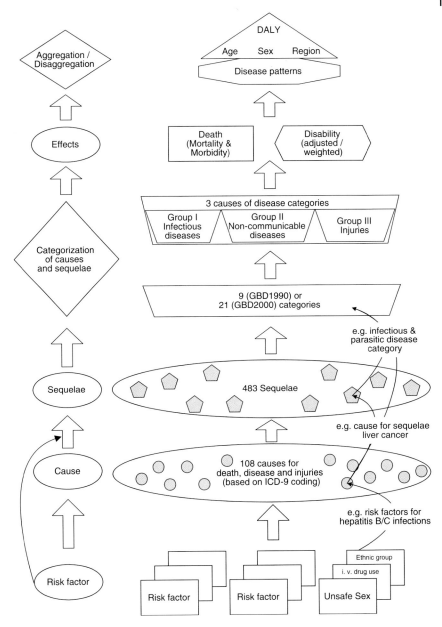

Figure 7
Procedure for determining DALYs.

of life as an infant. These weightings were arrived at through consensual procedures and set forth transparently.

By taking into account the factors of age and gender and discounting for time, one can transform the initially simple formulae for YLL and YLD

$$YLL = N \times L$$

where N = number of deaths and L = standard life expectancy for an age cohort, and

$$YLD = I \times DW \times L$$

where L = average duration of disability (measured in years), I = number of incident cases in the reference period, and DW = disability weight (from 0 = healthy to 1 = worst imaginable state of health), very rapidly into more complex equations, which will not be expanded upon within this book. It should simply be mentioned that they are dependent upon the discount rate, the age-weighting correction constant, and on other implicit assumptions (GBD 2001):

$$YLL = NCe^{(ra)} / (\beta+r)^2 \left\{ e^{-(\beta+r)(L+a)} \left[-(\beta+r)(L+a) - 1 \right] - e^{-(\beta+r)a} \left[-(\beta+r)a - 1 \right] \right\} \quad (3.1)$$

where YLL = years of life lost; N = numbers of deaths; C = age-weighting correction constant (GBD standard is 0.1658); r = discount rate (GBD standard is 0.03); a = age of onset; β = parameter of age-weighting function (GBD standard is 0.04); L = duration of disability or time lost due to premature death; and

$$YLD = I \times DW \times Ce^{(ra)} / (\beta+r)^2 \left\{ e^{-(\beta+r)(L+a)} \left[-(\beta+r)(L+a) - 1 \right] - e^{-(\beta+r)a} \left[-(\beta+r)a - 1 \right] \right\} \quad (3.2)$$

where YLD = years of life with disability; I = number of incident cases in the reference period; DW = disability weight (from 0 = no disability to 1 = worst imaginable state of health); C = age-weighting correction constant (GBD standard is 0.1658); r = discount rate (GBD standard is 0.03); a = age of onset; β = parameter of age-weighting function (GBD standard is 0.04); L = duration of disability.

Comparative risk evaluation
Within the GBD study, the term "comparative risk evaluation" signifies a systematic investigation of potential changes in the state of health of a population group as a result of a change in their exposure to a risk factor. Such changes are considered in relation to the effect of other risk factors.

To conduct such a comparative risk evaluation, the concept of the "attributable fraction" (AF) of the burden of a risk factor, for instance smoking, on a particular illness, for example lung cancer, was selected. This attributable fraction is defined as the reduction, expressed as a percentage, of the incidence of a disability or mortality caused by a risk factor when that factor is eliminated.

The following formula expresses this attributable fraction (AF):

$$AF = \frac{P(RR-1)}{P(RR-1)+1} \tag{3.3}$$

where P = prevalence of exposure; RR = relative risk.

Alternatively, the contribution of a risk factor can also be calculated by a comparison of the consequences of the exposure distribution (relating to this risk factor) with the consequences of a fictional, postulated distribution that matches a specified scenario, a counterfactual distribution.

For such a counterfactual distribution:

$$AF = \frac{\sum_{i=1}^{n} P_i \times RR_i - \sum_{i=1}^{n} P'_i \times RR_i}{\sum_{i=1}^{n} P_i \times RR_i} \tag{3.4}$$

where P_i = fraction of population in exposure category i; RR_i = relative risk for exposure category i: P'_i = fraction of population exposed to exposure category i in the counterfactual distribution; N = number of exposure categories or levels.

Should the distribution be continuous, for example in the case of high blood pressure in the population, or air pollution, then

$$AF = \frac{\int_{x=0}^{m} RR(x)P(x) - \int_{x=0}^{m} RR(x)P'(x)}{\int_{x=0}^{m} RR(x)P(x)} \tag{3.5}$$

where $P(x)$ = distribution of the population by exposure level x; $P'(x)$ = counterfactual distribution of exposure for the population; $RR(x)$ = relative risk of death or disease for exposure level x; m = maximum exposure level.

When one combines the attributable fraction of a risk factor for a specified cause of disease, injury, and death with the total burden for this cause, then one obtains the attributable burden for the risk factor:

$$AB = AF \times B \tag{3.6}$$

where AB = attributable burden; B = total burden for a cause.

With the help of these metrics various risk factors can be compared. Thus the most important risk factors can be identified and, if need be, reduced.

The complete procedure of a comparative risk evaluation consists of the following steps (GBD 2001, p. 121):
1. choice of risk factors;
2. choice of relevant diseases and injuries caused by the risk factor;
3. choice of appropriate exposure variable;
4. collection of data on population distribution of exposure for step 1;

5. specification of the risk factor–disease relationship for each disease and injury in step 2 due to exposure to the risk factor in step 1;
6. choice of counterfactual distribution of exposure;
7. calculation of the burden of disease or injury due to each cause in step 2;
8. analysis of uncertainty.

Steps 1–6 result, together with Eq. (3.4) and, as the case may be, Eq. (3.5), in the contribution of the risk factor in question to a specific disease or injury. From this and from the subsequent step 7, using Eq. (3.6), the burden of the risk factor is produced.

A well thought out and case-specific strategy must be developed for selecting risk factors, together with their causal relationships to particular causes of death, illness, or injury. Aspects to be considered are (Mathers et al. 2001):
- criteria for identifying relevant studies
- search strategy for studies
- characteristics of excluded studies
- description of studies, including methodological qualities
- characteristics of included studies such as sample size and region
- standard errors reported by studies
- steps to assess and reduce random error
- reported standard errors and other measures of uncertainty due to variability
- meta-analysis if appropriate
- steps to assess and reduce bias and to assess causality
- confounding
- selection bias
- information bias (exposure and outcome measurement error).

Figure 8 shows the significant input parameters and the results of one such comparative examination of a risk. Through the use of current or forecast distributions for a risk factor in comparison with counterfactual distributions, not only the influence of prior or current exposures on current burdens through disease but also the influence of current and future exposures on future burdens can be calculated.

Selection of affected regions and population groups
The approach to viewing the geopolitical aspects of the state of global health by examining global regions is rather arbitrary, but it does reflect traditional concepts of the development status of a global region. Nonetheless, it is clear that very different disease patterns may be found within a global region and even within a country. One need only consider, for instance, the difference between disease patterns in urban and rural areas. Other influencing factors stem from purely social differences, such as religion, ethnicity, or cultural customs. However, this heterogeneity suffers from the global treatment of statistics because of the high level of aggregation demanded; even within the region-specific disaggregation, this heterogeneity can only be inadequately observed. It is for this reason that the WHO developed in-

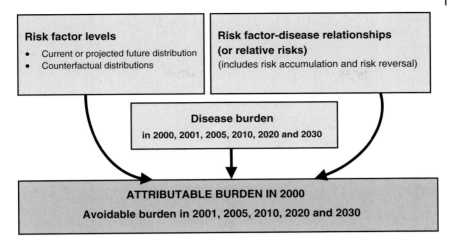

Figure 8
Input/output for GBD comparative examination of risk (GBD 2001).

struments to support the implementation of national and subnational health burden studies.

Selection of timeframes to be considered and choice of discounting
The selection of timeframes in the GBD projects is of fundamental significance. This is not only due to the use of time as a yardstick and the arbitrary choice of how long a standard life expectancy should be that this involves, but also the issue of how an age cohort should be defined, especially in regard to assumptions about changing and time-dynamic mortality rates.

Uncertainties
Few uncertainties were discussed in the first study (GBD 1996): the discount rate and the age weighting have a notable influence on the results, while the difference between the weightings for various disabilities in the DALY metric is narrow. The (unquantified) uncertainties of the epidemiological data influence the results significantly more strongly (Murray and Lopez 1996b). In GBD (2000) the uncertainties of data are considered (Mathers et al. 2002), as are the uncertainties of the DALY metric's weighting (Salomon et al. 2001).

Stakeholders
The first GBD study was conducted as a widespread collaboration between more than 100 scientists in 20 nations. The following discussion, however, indicated that there were gaps in the pattern of participation. Thus not only were representatives of certain expert fields absent, but also stakeholders from outside the expert community. This last absence is particularly problematic because the appraisal of desired values for the fundamental criteria of the state of health of a population, for

instance the measurement of a preference for a state of health through the affected person, is closely coupled with societal values. The operationalization of such societal values demands the involvement of a broader spectrum of stakeholders. The WHO's strategy of providing instruments to support health burden studies is an important step in this direction.

Conclusion

The comparative examination of risk factors affecting a health burden was, in the GBD study, performed by means of a highly aggregated burden metric (DALY), which involves the integration of disease patterns. Although the DALY metric represents an intuitive and simplified value dimension, its evaluation (i.e., calculation) is in practice a highly complicated undertaking which can in most cases only be accomplished with specially developed software.

The gain or loss of years of life is provided as a measure of the significance of risk factors. What is notable in the GBD project is that such measure of risk should be defined with careful thought and presented transparently.

Table 3 gives an overview of the procedure.

Table 3 Procedure in the Global Burden of Disease (GBD) study.

Objective	Consistent evaluation of the main causes of cases of death and diseases leading to disability, and cases of death and disability for 10 significant risk factors as well as projections for mortality and disabilities Establishment of priorities for public health policy
Spheres of risk	Group I: infectious and parasitic diseases, influences before and during pregnancy, malnutrition Group II: noncommunicable diseases Group III: injuries (intentional and unintentional) A list of various diseases, and causes and sequelae, was developed through a tree structure
Comparison categories	Values for various calendar years
Timeframes	1 calendar year each
Geographic frame	Global and divided into 8 geographical regions
Attributes	Cases of death, disease, disability
Aggregation/metric	DALY; weightings arrived at through surveys of small groups, differentiation of weightings between various countries only slight
Overall evaluation	Implicitly multiattribute
Uncertainties	In GBD1996 some uncertainties discussed qualitatively, but without quantitative treatment; this is dealt with in GBD2000
Presentation of results	Tables and graphs of both total cases and in the units of the metric (DALYs)

3.2.4
ExternE Project

Procedure
The ExternE project (ExternE Summary 1995) is presented here as a comparative risk evaluation of energy generation options. Together with the typical procedures and outcomes of the project come a typical approach and the typical open questions.

Significant advances were made by the consistent and comprehensive determination of the external costs of energy sources through the ExternE project. The project enabled a comparison of energy sources. The project
- considered the whole lifecycle of a technology;
- examined the direct consequences of the emissions and burdens of a technology (in principle, at least: limited knowledge led to some limitations), but no indirect consequences, e.g., effects occurring in the workplace;
- considered the consequences for future generations;
- was based on the effect-path procedure (Figure 9);
- calculated all damages on a unified scale (in this case, financial value) to enable comparisons to be made.

What is important about this risk comparison is that the comparison does not take place through a consideration of the results obtained for various energy systems, but that the comparison is already incorporated within the calculation of results. Thus through the choice of measuring units, for instance, one can determine how a case of sudden death is evaluated against a case of cancer appearing after a long period of latency, how the illness resulting from such a case, whether transient or permanent but not resulting in death, is evaluated, and how all of these are compared to, for example, ecological damage or damage to intangible goods. Other terms, some normative and some arbitrary, have a strong influence on the external costs calculated, and thus also on the findings of such a comparison.

The procedure is elucidated below. The problem assumptions are discussed for those phases of the analysis in which they are used.

Aim of the project
The aim of the project is the determination of the external costs of energy production. It is justified as follows:

> *In Europe, policy analysts are being required to take account of environmental aspects in their decision making and to undertake cost–benefit analysis of available options. The 5th Environmental Action Programme Towards Sustainability, clearly indicates the need for the assessment of externalities and monetary evaluation.*
> (ExternE Overview, undated)

Model

The model is based on the analysis of "damage function" or "effect path". In ExternE Overview this is illustrated in a diagram in which an important step is missing: the aggregation of damages. Figure 9 is based on this diagram, but is completed by the addition of the aggregation of damages.

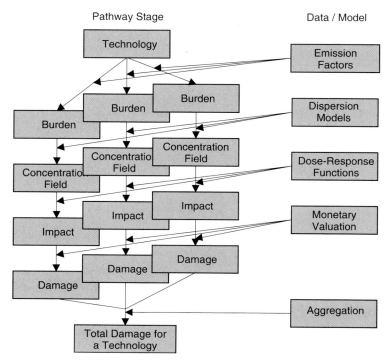

Figure 9
Typical impact pathway approach. (From ExternE Overview, with the addition of the aggregation of several impact paths.)

Selection of technology

As already mentioned, the project examines energy production. Accordingly, only energy production technologies which are currently in operation are considered (fossil: anthracites, lignites, oil, gas, nuclear power; renewable: hydroelectric, photovoltaic, wind, biomass; also energy conservation) (ExternE Summary 1995, p. 12).

Selection of subjects of protection, influences, and effects considered

This proceeds from the known harmful effects of substances known to be released by the types of energy production under consideration. In principle, all phases of the fuel cycle (e.g., primary extraction, disposal of spoil, fuel preparation, transport, combustion, disposal of combustion products) and all phases in the lifecycles of the necessary equipment (installation, operation, disassembly, disposal) are examined.

In ExternE Overview the effects considered are described as follows:

> *External costs of energy are ... costs for society and the environment which are not considered by the producers and consumers of energy, i.e., which are not contained within the market price. These include damage to the natural and built environment, such as*

- effects of air pollution on
  - health
  - buildings
  - agricultural incomes
  - forests
  - global warming
- occupational disease and injury
- adverse effects of a plant's visual aspect or of noise emissions.

Further are mentioned:
- freshwater fisheries
- biodiversity.

On the basis of the first rough assessments, only those causes and effects that contributed significantly to calculated damages were included in the actual quantification. It was emphasized throughout, however, that all causes and effects had in principle been examined (ExternE Methodology Annexes 1997).

Selection of affected regions and population groups
The calculations were performed for specific locations. This is because, just as the concentration of harmful substances decreases with increasing distance from the facility discharging them, so the damages are greater the higher the population density in the vicinity of the facility in question. Likewise, fewer damages were designated for coastal locations than for those inland because the proportion of discharged pollutants carried out to sea can cause no harm to health, there being no population there (Mayerhofer et al. 1997, p. 105).

> *The system boundary also has spatial and temporal dimensions. Both should be designed to capture impacts as fully as possible.*
>
> *This has major implications for the analysis of the effects of air pollution in particular. It necessitates extension of the analysis to a distance of hundreds of kilometres for many air pollutants operating at the "regional" scale, such as ozone, secondary particles, and SO_2. For greenhouse gases the appropriate range for the analysis is obviously global. Consideration of these ranges is in marked contrast to the standard procedure employed in environmental impact assessment which considers pollutant transport over a distance of only a few kilometres and is further restricted to primary pollutants. The importance of this issue in*

> *externalities analysis is that in many cases in the ExternE project it has been found that regional effects of air pollutants like SO_2, NO_x and associated secondary pollutants are far greater than effects on the local scale.* (Mayerhofer et al. 1997, pp. 9–10)

This is tantamount to the exclusive examination of collective risk. The individual risk caused by a plant to an individual living nearby is, naturally, independent of the number of people living near the plant.

Selection of timeframes to be considered and choice of discounting
In principle all periods in which effects are anticipated are considered:

> *Despite the uncertainties involved it is informative to conduct analysis of impacts that take effect over periods of many years. By doing so it is at least possible to gain some idea of how important these effects might be in comparison to effects experienced over shorter time scales. The chief methodological and ethical issues that need to be addressed can also be identified. To ignore them would suggest that they are unlikely to be of any importance.* (Mayerhofer et al. 1997, pp. 10–11)

> *...the collective dose resulting from uranium mining and milling ... is mainly caused by the release of radon from abandoned tailing piles assumed to continue over 10,000 years ... Because of the time horizon considered, the discounted present value of the damage is rather small, while the undiscounted damage from the uranium mining and milling stage dominates the total external costs of the fuel cycle.* (Krewitt et al. 1997, p. 80)

Nonetheless, significant uncertainties remain:
- The choice of a period of 10,000 years is arbitrary. It has by no means been established whether the adverse effects will have faded to negligibility. Rather, according to Krewitt et al. (1997, p. 38): "...an exhalation rate of 3 $Bq/m^2/s$ from reasonably covered mill tailings that is expected to remain unchanged over at least 10,000 years". If one were to calculate for a longer period, the estimate damages would be higher, and for a shorter period, lower.
- The long-term effects will occur in a future in which we cannot determine how they will appear:

> *These impacts will be imposed on societies with very different characteristics from those of today. The magnitude of the impact can vary by orders of magnitude depending upon the scenario assumptions. These uncertainties are far larger than those caused by scientific uncertainty.* (ExternE Summary 1995, p. 153)

Basing calculations on such long periods clearly has a decisive influence on whether and to what extent future damages should be discounted (ExternE Summary 1995, pp. 43–46). Neither is this a question that can be answered scientifically, but an evaluative issue.

Furthermore, the consequences of uncommon accidents are averaged over time (ExternE Methodology Annexes 1997, p. 34; Krewitt et al. 1997, ES-4, p. 38). However, if one proceeds from the fact that the worst category of nuclear accident would entail the total economic destruction of a country, then averaging the consequences of such an accident over time makes no sense.

Dose–effect relationships
The calculations in the ExternE project are based on known and acknowledged dose–effect relationships. Nonetheless, ascertaining the consequences even of known burdens is difficult. The following two problematic aspects were identified within the project.

The first aspect is that some pollutants have a threshold value and/or a nonlinear dose–effect relationship, so that the effect of the additional emissions from the energy production system under examination is dependent upon the prior emissions of other systems. The second aspect is that the impact of a pollutant can depend on the concentration of other substances that are produced for the most part not by the energy system being studied but by other systems. Thus the damage caused by the energy production system depends on how high the emissions of other systems are:

> *One of the main findings concerning the aggregation methodology is that damage costs per ton of pollutant emitted might differ significantly even within a country. The main reason for such differences is the spatial variation in SO_2, NO_x, and NH_3 emissions. As ammonia is mainly emitted from agricultural activities, and the availability of free ammonia in the atmosphere also depends on the level of SO_2 and NO_x emissions from other sources, the damage costs we refer to as "energy" externalities in fact strongly depend on emissions from various industrial activities.* (Krewitt et al. 1997, ES-6; p. 76)

The question left open to the decision-maker then concerns to which polluter to ascribe such multicausal damage.

In any case, the prior burden is considered. The additional damage caused by an additional facility is calculated from the difference between the dose–effect relationship of the previous burden and that of the total burden including that of the additional facility. Thus a situation arises in which pollutants, which do have a dose–effect relationship exhibiting a threshold value, present a differential increase of damages as though they possessed an (approximately linear) dose–effect relationship without threshold value. This leads to an identical rise in damage to health, regardless of whether 100 individuals are burdened with a 10% increase in pollutant or whether 100,000 individuals are burdened with an increase of 0.01%. Whether such an equivalence can be justified is indeed questionable.

In addition, the estimated damages depend on the order of examination, which – depending on the problem under consideration – may be predetermined or arbitrary. If one begins with an unburdened situation and adds one polluter after the other, no damages arise for the first polluter considered, so long as the burden remains below the threshold value. Damages only arise for the polluter under consideration when the threshold value is crossed.

Finally, the project expressly determines the *incremental* consequences of, for instance, a new power plant project or changes in energy policy (ExternE Methodology 1995, p. 11). This means that the values arrived at become less reliable the more the actual total burden deviates from the situation considered in the project.

Valuation of damages
To enable a comparison of different energy systems, the damages are transferred to a unified comparative scale. In ExternE, this is a financial value. The following problems, however, also occur when a scale of measurement other than financial value is chosen.

The evaluation of immaterial damages
How does one evaluate, for example, the value of a forest as a recreational area? "The analyst should collect information for the decision maker and not make arbitrary decisions about which aspects are worthy of further discussion".

> Many receptors that may be affected by fuel chain activities are valued in a number of different ways. For example, forests are valued not just for the timber that they produce, but also for providing recreational resources, habitats for wildlife, their interactions (direct and indirect) with climate and the hydrological cycle, protection of buildings and people in areas subject to avalanche, etc. Externalities analysis should include all such aspects in its valuation. Again, the fact that a full quantitative valuation along these lines is rarely possible is besides the point when seeking to define what a study should seek to address: the analyst has the responsibility of gathering information on behalf of decision makers and should not make arbitrary decisions as to what may be worthy of further debate.
> (Mayerhofer et al. 1997, p. 9)

The evaluation of various cases of death
The first phase of the ExternE project began with the "value of a statistical life" (VSL). However, are all cases of death to be equally valued, whatever the age and under whatever ancillary conditions the affected individual died? This is open to question, particularly in cases of cancer occurring only at high age, or in cases of death through air pollution that in most cases typically strikes only people who are already in a weakened state of health and who quite possibly would not have lived for much longer without this anyway.

The discussion of the methodology introduces the dimension of "years of life lost" (YOLL). It is explained:

> ...in earlier phases of the project a number of questions were raised regarding the use of the VSL for every case of mortality considered. These originally related to the fact that many people whose deaths were linked to air pollution were suspected of having only a short life expectancy even in the absence of air pollution. Was it logical to ascribe the same value to someone with a day to live as someone with tens of years of remaining life expectancy? Furthermore, is it logical to ascribe the full VSL to cases where air pollution is only one factor of perhaps several that determines the time of death, with air pollution playing perhaps only a minor role in the timing of mortality? In view of this the project team explored valuation on the basis of life years lost.
>
> ...Within ExternE it has been concluded that VSL estimates should be restricted to valuing fatal accidents, mortality impacts in climate change modeling, and similar cases where the impact is sudden and where the affected population is similar to the general population for which the VSL applies. The view of the project team is that the VSL should not be used in cases where the hazard has a significant latency period before impact, or where the probability of survival after impact is altered over a prolonged period. In such cases the value of life years (YOLL) lost approach is recommended. (Mayerhofer et al. 1997, p. 260)

The choice between one measure (VSL) and the other (YOLL) itself constitutes a decision for a specific evaluation of cases of death and against the other.

The evaluation of various diseases in relation to their duration and gravity
Similar problems arise here as in the evaluation of different cases of death: the scale, using time of onset and duration of disease, involves a certain evaluation of diseases. A further evaluative scale specifies how the gravity of the disease is incorporated.

The financial valuation of health and environmental damages and cases of death
This procedure often attracts criticism because it sets life and health against material values. (This criticism, however, is drawn by any comparison of differing types of harm, as will be discussed later.)

> Although valuation of environmental impacts in monetary terms is widespread, there are still many people who find the idea strange at best, and distasteful and unacceptable at worst ... The essence of economic valuation of environmental goods is that it attempts to identify the preference of individuals for the allocation of resources by, for example, quantifying people's "Willingness to Pay" (WTP) for environmental improvement. It does not pretend to quantify the intrinsic value of a human life or biodiversity ...

> It is, however, acknowledged that there are ranges of problems
> with the use of valuation on the base of WTP:
> 1. The concept is "income constrained". Since you cannot pay
>   what you do not have, a poorer person's WTP is less than that
>   of a richer person, other things being equal ...
> 2. Putting money values on health and the environment and
>   pricing accordingly does not necessarily guarantee
>   sustainability, and therefore taken alone it is insufficient as a
>   prescription for economic policy on the environment.
> 3. At the most fundamental level, it is disputed whether
>   monetary valuation of health and the environment is
>   philosophically defensible. This issue is outside the scope of our
>   study.   (ExternE Summary 1995, pp. 41–42)

Regardless of whether a unified scale of damages uses financial valuation or some other measure, such a scale must of necessity compare heterogeneous forms of harm, be they loss of human life, adverse effects on health or on the environment, and, perforce, upon immaterial goods. This could only be avoided by abandoning the use of a unified scale and instead showing the results for the various damages separately. In the end, however, if one wants to compare risks, one cannot avoid weighing them against each other.

Uncertainties

Uncertainties are examined systematically, as far as possible. Statistical uncertainties, uncertainties arising from the model, uncertain assumptions, and errors were distinguished (ExternE Methodology Annexes 1997, p. 87):

> It is appropriate to group the main contributions to the
> uncertainty into qualitatively different categories:
>
> - *statistical uncertainty* – deriving from technical and scientific
>   studies, e.g., dose–response functions and results of valuation
>   studies;
> - *model uncertainty* – deriving from judgments about which
>   models are the best to use, processes and areas excluded from
>   them, extension of them to issues for which they are not
>   calibrated or designed. Obvious examples are the use of models
>   with and without thresholds, use of rural models for urban
>   areas, neglecting areas outside dispersion models and transfer
>   of dose-response and valuation results to other countries;
> - *uncertainty due to policy and ethical choices* – deriving from
>   essentially arbitrary decisions about contentious social,
>   economic and political questions, for example decisions on
>   discount rate and how to aggregate damages to population
>   groups with different incomes and preferences;
> - *uncertainty about the future* – deriving from assumptions
>   which have to be made about future underlying trends in

health, environmental protection, economic and social development, which affect damage calculations, e.g., the potential for reducing crop losses by the development of more resistant species;
- human error.

For human error, little can be done other than attempting to minimize it. The ExternE project uses well-reviewed results and models wherever available and calculations are checked. The use of standardized software (EcoSense) has greatly assisted this. Uncertainties of the first type (statistical) are amenable to analysis treatment by statistical methods, allowing the calculation of formal confidence intervals around a best estimate. Uncertainties in the other categories are not amenable to this approach, because there is no sensible way of attaching probabilities to judgments, scenarios of the future, the "correctness" of ethical choices or the chances of error. There is no reason to expect that a statistical distribution has any meaning when attempting to take into account the possible variability in these parameters. In addition, our best estimate in these cases may not be a median value, thus the uncertainty induced may be systematic. Nevertheless the uncertainty associated with these issues is important and needs to be addressed.

The uncertainties of the results were derived from the uncertainties of the parameters using a simple arithmetical method. This method assumes that the results are largely calculated through the multiplication of parameters.[38] When the uncertainty of each parameter x_i $(i = 1\ldots n)$ is approximated through a log normal distribution using the standard geometric deviation, the uncertainty of the product of this parameter can be calculated using the formula

$$\left[\log(s_{g,\text{prod}})\right]^2 = \prod_{i=1}^{n}\left[\log(s_{g,i})\right]^2 \tag{3.7}$$

This formula does not work for the uncertainty of the sum of uncertain values. Hence no uncertainties are given for the aggregated (summed) risks. The uncertainties were, however, examined through a cost–benefit comparison; the effects (attributes) were taken into consideration in order of "confidence attached" and compared with the costs of their abatement thus (Krewitt et al. 1997, pp. 89-90; ExternE Vol. 10: National Implementation 1999, p. 153):

> In order for abatement technology to be considered worthwhile it is necessary for benefits to exceed costs. Comparison of the private costs with the estimated damage costs per ton of pollutant emitted suggests that the abatement measures reviewed by ERM

[38] For example, Damage = pollution concentration × population × exposure–response function × valuation (ExternE Methodology Annexes 1997, p. 88).

(1996) are justified for both UK and German plant. However, the externality estimates are subject to significant uncertainty. To take into account these uncertainties, we rank the results for each impact of the three pollutants in terms of the confidence attached to each (see, e.g., Figure 9.7 [Figure 10]). For the German plant, current best available emissions control techniques as defined in ERM (1996) is justified without reference to mortality effects. In fact for SO_2 it is only effects on materials and crops that need to be included before the lower end of the range of private costs is passed. For the UK plant, where overall damages are lower because of the location of the power plant site close to the North Sea, the combined effect of damage to materials and crops (where appropriate) and acute morbidity are sufficient to take results for all three pollutants past the lower end of the range for the private costs of abatement, and also past the upper end of the range for particulates.

Stakeholders
Over 30 teams from research institutes and consulting firms from nine member states of the EU and other European countries took part in the project. Whether, and which, stakeholders took part was not clear from the available information.

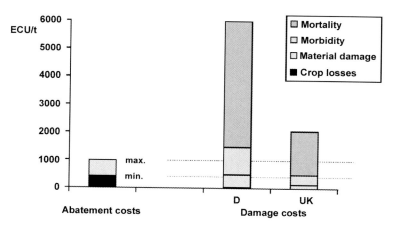

Figure 10
Comparison of abatement costs and damage costs per ton of SO_2 emitted from a coal-fired power plant in Germany and the UK.
(From Krewitt et al. (1997).)

Conclusion

Strictly speaking, the project dealt with external costs rather than risks. However, since a great proportion of external costs result from risks, it is possible to draw many conclusions for risk comparisons from this project. It should, nonetheless, be observed that the aim of the project is not a completed comparison, but the delivery of data and techniques to enable comparisons to be made. Because the external costs depend in particular on the population density around the location of a plant and from the burden of pollutants from other sources, such comparisons are specific to locations and so can only be meaningfully used for defined scenarios.

A particular merit of the project is that it documented assumptions, which are often made without comment and discussed them to some extent. The following deserve special mention:

- The financial valuation of immaterial subjects of protection provokes contention, both on principle (i.e., immaterial subjects of protection should never be equated with material goods) and in respect of the weighting (how much money an ecosystem is actually worth).
- The effect of a marginal dose may be overestimated. On the basis of a marginally higher radon burden a significantly higher number of cancer cases is detected because a (arbitrarily chosen) selected period of 10,000 years and the number of people living in this period of time are aggregated. If this period were reduced to 100 years (or future damage were discounted), this burden of pollution would no longer be significant.
- Grave yet very infrequent accidents were averaged over time. (Only the product "frequency of incidence of damage" is included in the result.)
- The discounting of future damage is contentious. Therefore results were delivered for various discount rates.
- An innovation in the process of examining the uncertainty of results is not to aggregate the single contributions directly but to begin with those with "most confidence attached" and proceed in descending order until a conclusion is possible. Thus many relatively certain conclusions become possible without having to revert to the more uncertain contributions.

Table 4 provides an overview of the procedure.

Table 4 Procedure of ExternE project.

Objective	Comparison of alternative technologies; "public health", "public wealth"
Spheres of risk	Risks arising from electricity generating plants
Timeframes	In principle unlimited (10,000 years for radon); future damage was discounted
Geographic frame	Only limited by the area polluted. Regional peculiarities were considered
Attributes	Damage to health, buildings, agricultural and fisheries proceeds, forests, and biodiversity, and through global warming; occupational illnesses and accidents in the workplace; adverse effects through appearance of a facility or noise pollution
Aggregation/metric	VSL or YOLL, financial evaluation
Overall evaluation	Comparison of aggregated values; implicitly multiattribute through the metrics applied
Problematic assumptions	Financial valuation of intangible subjects of protection Aggregation of damage caused by extremely low doses in extremely large populations Evaluation of extremely rare large-scale accidents only through the product "frequency of incidence of damage" Discount rate for future damage
Uncertainties	Statistical uncertainties systematically examined and presented in detailed tables of results. Within the comparison the contributions were put in order of "confidence attached". Sensitivity analyses of some (but not all) uncertain assumptions were performed
Presentation of results	Detailed tables for every technology, broken down by attributes; no overall comparison, but examples of application in cost–benefit comparisons, along with examination of uncertainties

3.2.5
Comprehensive Assessment of Energy Systems Project

The Comprehensive Assessment of Energy Systems (Ganzheitliche Betrachtung von Energiesystemen or GaBE) project, which has been run for several years by the Paul Scherrer Institut (PSI), belongs to the field of risk, cost, and benefit analysis, with the result of a choice between options (Hirschberg and Voss 1999). It therefore addresses the third goal mentioned in Chapter 1 of this book. The project has as its aim the development of a comprehensive and useable methodology for the consistent and detailed evaluation of energy systems. In this, sustainability was not originally a research topic, but over the course of time a framework concept for the evaluation of energy systems was developed from this approach, not least through the adoption of sustainable development as a guiding principle within the Swiss constitution.

A forerunner: the Poly project
As early as the beginning of the 1990s, the Poly project at the ETH in Zurich had taken up the theme of risk comparison (Berg et al. 1995). An important goal of this project was to develop contributions towards a method of assessing risk arising from technical systems that could be generalized. The Swiss statutory order on hazardous incidents (Schweizer Störfallverordnung, StFV) was selected as a basis for the development of a suitable risk evaluation in the energy field. With the instrument developed for this order, it was attempted to record and evaluate the risks associated with hazardous incidents arising from processing chemical substances and microorganisms in stationary facilities as well as from the transport of hazardous goods. The order does not extend to risks arising from the operation of energy generating – or, more precisely, energy converting – systems. Because it is, moreover, known that the regular, i.e., incident-free, operation of these systems also has significant effects on the environment, three questions were addressed (Berg et al. 1995):

- Is the StFV's approach to risk suitable for the evaluation of environmental risks in technical systems?
- Can the StFV's methodology be applied to hazardous incidents in the energy field?
- Can the terminology and methods of risk science, which are oriented towards hazardous incidents, be transferred to risk-related aspects of normal operation?

The StFV's approach to risk focuses on the identification of hazardous incident scenarios to which a risk, in the form of a combination of the extent of damage and the probability of occurrence, is assigned. Different methods of scenario development are used for this, each according to the system being investigated. Through their analysis the authors (Berg et al. 1995) reach the conclusion that the environment of the facilities analyzed, understood as a system, is too complex to allow the definition of endpoints needed for a damage analysis or the formulation of cause–effect relationships. Further, they assert that the StFV's approach to risk only covers the operation of a facility and should be augmented by a lifecycle assessment. The wider the framework of the examination, according to the authors, the more likely it is that the approach makes accessible only a part of the aspects that must be considered in a comprehensive evaluation.

The elaboration of risk evaluation in the energy field mentioned (Berg et al. 1995) contains a compilation of the environmental insults brought about in the course of the extraction, transport, and preparation of primary energy sources, the supply and final use of energy, and the disposal of waste. Among these environmental insults, the creation of the potential for hazardous incidents and accidents can be included. An assessment of the effects of these insults upon the environment has found that long-term changes to the environment are as a rule the consequences of regular operational processes, whereas hazardous incidents could be particularly conspicuous and locally serious. It is only in the cases of nuclear power and the transport of mineral oil that environmental effects are not clearly dominated by regular operation.

Specific difficulties are encountered in the application of the StFV's approach to risk to normal operation and hazardous incidents. With hazardous incidents the question arises of the relevance of examinations of frequency and probability. Hazardous incidents with great magnitudes of damage normally have very low calculated probabilities of occurrence. These are difficult to determine in a convincing way. In such cases the magnitude of damage is treated as the dominant evaluative factor. The risks of normal operation are, in contrast, caused by constant effects that mostly contribute to an already existing burden. Here it is difficult to establish a causal connection between emission (from a specific facility) and effect.

The authors (Berg et al. 1995) propose the replacement of damage as the basis of assessment with environmental hazard. They choose indicators, which, on the level of exposures, examine three key parameters for the environmental hazard: spatial and temporal range, and magnitude. As a further criterion they bring to bear the issue of whether the environmental effect is the object of current public debate. This leads to the use of CO_2 emissions, radioactive wastes, and demands on land as indicators. The calculation of numerical values proceeds on the basis of an elaboration to ecological inventories of energy systems (Frischknecht et al. 1996). In these ecological inventories very rare hazardous incidents below a threshold frequency – oriented on events per energy conversion – are not examined; this includes nuclear accidents and large oil tanker catastrophes. What is characteristic of the analysis is that their sequence turns out differently depending on indicator. However, a scale for the "total hazard" is not available in Berg et al. (1995).

What is significant here is that the aggregation of single aspects to form a unified expression on the level of risk comparison is difficult. Such questions of aggregation are also a theme of the GaBE project.

The GaBE project
The GaBE project (Hirschberg et al. 1998; Hirschberg and Voss 1999; Kröger et al. 2001) takes up elements of the Poly project; it analyses and evaluates, in an interdisciplinary manner and following an integral concept, the following factors as regards energy supply:
- environmental influence of normal operation
- health effects of normal operation
- risks of severe hazardous incidents (severe accidents)
- economic aspects.

Principally, the methodology developed within the framework of the PSI's project allowed the following questions to be investigated:
- In which respect can a specific technique be improved (comparison of nonaggregated indicators)?

- How can several technological options (contemporary/future facilities) be compared, for instance with specific indicators or with aggregated values?
- What does an efficient comparison of different scenarios of energy supply look like?

The project explicitly pursued the aim of ascertaining risks for the purposes of comparison, in which issues about the choice between various energy supply options played a significant role.

Approach and procedure
A central component of the methodology applied is the life cycle assessment (LCA), which takes comprehensive account of the normal operation of facilities (Frischknecht et al. 1996). Kröger (1998, 2000) proposes, in the context of the GaBE project, an evaluative matrix ("framework") for energy systems that explicitly uses a quantitative procedure oriented on the "three pillars concept" of sustainability. Three principles (avoidance of exhaustion of resources, avoidance of non-biodegradable wastes, and sensitivity to the environment) combine ecological, economic, and social aspects and are concretized through criteria and indicators (see Table 5). Decisive characteristics of this approach are its quantifiability and the appropriate units of measurement of the indicators ("sustainability is measurable"). Thus aspects that are usually investigated when dealing with environmental risks are addressed in the criteria category of "no depletion of resources" (see below).

The applied indicators permit the inclusion of both the present and advanced future systems of electrical energy supply. As comparisons indicate (Kröger et al. 2001), these indicators also cover the relevant attributes suggested for the evaluation of energy systems suggested by institutions such as OECD/NEA and UN/IAEA. Work at PSI investigated whether a databank allows the reliable quantification of indicators by means of a comparison of various energy chains (Kröger et al. 2001). It also investigated whether such an assessment can demonstrate the fundamental differences between energy supply options to the general public in a scientifically irreproachable manner. A suitable data table was published by PSI in the run-up to a referendum in Switzerland on renewable energy. It contains indicators for conservation systems based on the energy sources coal, natural gas, nuclear fission, hydroelectric, and photovoltaic. In Table 6 (Kröger et al. 2001), emission of SO compounds are also given; they are representative of negative effects on health. The authors are explicit in making the sources of data and their attendant assumptions transparent, because even apparently minor changes may have a significant influence on the results.

Treatment of risks
The GaBE project confers a particular significance on the topic of risk. In doing so it deals with environmental risks; they are considered from a sustainability-oriented standpoint in which appropriate subjects of protection are understood as resources. The term resources is to be understood here in a broad sense and encompasses the health of people and nature, the economy, and social peace. Risks are

Table 5 Concretization of the concept of sustainability in relation to energy systems. (After Kröger (2000).)

Criteria categories	Specific criteria	Indicators		Units of measurement
"No" depletion of resources	Use of fuel and materials	Period of degradation		Years
	Loss of land	Unusable area × duration	Operation	km² × years
			Severe accidents	km² × years
	Environmental effects		Greenhouse gases	Tons CO_2 equiv.
			Acid rain	Tons SO_2 equiv.
	Effects on health	Acute and late deaths	Operation	Number of deaths
			Severe accidents	Number of deaths
			Objective risk evaluation	Number of deaths
	Social aspects	Acceptance		Δ %
		Risk of proliferation		Qualitative
		Economic competitiveness		Unit of currency/kWh
"No" production of nonbiodegradable wastes		Amount of waste		m³ or tons
		Confinement times		Years
High potential for robustness, long-term stability, less sensitivity to external conditions	Supply security	Foreign dependence		Qualitative
		Availability of technology		Unit of currency
	Robustness	Lag time		Hours
		Sociopolitical conditions		Qualitative

thereby understood as consequences of possible hazardous incidents or accidents – particularly serious accidents – as a special and important aspect of the evaluation of energy systems; the consequences for health ("health risks") of normal operation are conceptually separated from these risks arising from hazardous incidents and are a component of the analyses in the course of a "classic" LCA.

Risks due to serious hazardous incidents were included in a methodological enhancement of the LCA within the GaBE project (Hirschberg et al. 1998; Kröger et al. 2001) by an appraisal of reports concerning past events or by means of prog-

Table 6 Selected (exemplary) indicators for present and future energy systems (Kröger et al. 2001).

Indicator	Unit	Hard coal	Natural gas	Nuclear	Hydro (storage)	Photovoltaic
Fuel reserves	Years	160–2 300	70–170	120–400	–	–
Material consumption (copper ore)	kg/GWh$_e$	14–19 59	16 8	7–9 4	<1 <1	270–1600 350
Greenhouse gases	t(CO$_2$-equiv)/GWh$_e$	950–1200 770	530 390	8–29 6	4 4	110–260 44
Sulfur dioxide	kg(SO$_x$)/GWh$_e$	920–25,000 520	260 150	56–150 33	8–10 7	700–3600 160
Inorganic waste in repository	kg/GWh$_e$	5800–54,000 4000	1500 1100	650–1200 600	30 30	4900–10,000 1600
Highly and moderately radioactive waste	m^3/GWh$_e$	0.13–0.20 0.04	0.04 0.004	9.0–11.0 2.4	0.006 0.002	0.6–1.2 0.06
Production costs	Rp/kWh$_e$	5.7–7.4 6.3	4.7–5.8 4.7–8.2	5.1–7.5 5.7–7.2	4–21 12–16	70–140 45
External (environmental) costs	Rp/kWh$_e$	3.1–15.8 5.1–8.6	0.8–5.5 2.5–4.2	0.2–1.3 0.3–0.4	0–1.2 0.1	0.1–1.5 0.5–0.7

The upper figures represent bandwidths of values for present-day systems; the lower figures contain the corresponding values for the best future systems (specifically Swiss constraints).

nostic methods, for instance probabilistic risk analyses. A broad spectrum of damage is included (death, serious injury, evacuation, contamination of earth and water, and financial loss). An aversion to harmful events of catastrophic magnitude is measured by loss of useable land and number of deaths.

Among the findings of the GaBE project is the important conclusion that a comparison between various energy chains is hampered by the fact that statistical data exist for the calculation of the risks of conventional energy chains, whereas in the case of nuclear power – at least, for western reactors – the results of probabilistic risk analyses, i.e., of prognostic methods, must be applied. Moreover, statistical data are not without problems either; historical data demand a careful examination of consistency and applicability.

Aggregation of single aspects
Ideas and some initial experience exist for the weighting and aggregation of the indicators proposed and their eventual use in decision-making processes. However, a workable plan has yet to be developed (Kröger et al. 2001). Some results have been used by energy supply companies in Switzerland for the development of strategy and in public debates (the referendum on the energy policy question).

One possibility for aggregation is through the calculation of costs. Thus the full costs, which contain both those of the established components (internal costs) and also the costs of influences on the environment (external costs), can be used as a measure of the resource-intensiveness of energy systems. Environmental effects, including harm to health, are calculated as an external cost. External costs are used as a highly aggregated indicator of influences on the environment; the total costs (internal and external) are viewed as an indicator of sustainability. In recent years, according to Hirschberg and Voss (1999), great advances have been made in the calculation of external costs. Even when the financial valuation of the consequences of the greenhouse effect or of damage to an ecosystem is not yet possible, the principal advantage of such an approach lies in the use of empirically ascertained individual preferences which "are as a rule preferable for the evaluations of individual decision makers or groups" (Hirschberg and Voss 1999).

Another approach to aggregation is based on multicriteria decision-making procedures that also, characteristically for discussions of sustainability, take account of the social dimension. Table 7 shows the evaluative criteria used by Hirschberg and Voss (1999) and their relative weightings.

The aim here was the comparative evaluation of the environmental health effects of various power generation systems (natural gas, hydroelectric, wind, nuclear, and photovoltaic power); in other words, in the sense in which the term is used in this book, a pure risk comparison. The weightings given are for demonstration purposes and can be case-specifically chosen. The quantification of the single criteria stems from LCA and from special considerations of risks arising from severe hazardous incidents. In use, the (as mentioned demonstrative) weighting factors gave the best aggregated values for wind energy and hydroelectric power. By way of trial, the weighting of the three main criteria was altered in the multicriteria assessment. Among other things, this showed in particular that the evaluation of nuclear

Table 7 Evaluative criteria and relative weightings for damage to the environment and health (Hirschberg and Voss 1999).

Main criterion and weightings	Second-level criterion and weightings	Third-level criterion and weightings
Burden arising from normal operation; 50	Local; 10	CO; 40 Noise; 40 Adverse effect on the landscape's appearance; 20
	Regional; 30	SO_x; 30 NO_x; 40 NMVOC; 10 Particulate matter; 15 Radiation; 5
	Global (CO_2 equivalent); 60	
Burden arising from severe accidents; 25	Environment; 25	Contamination of land and inland water; 70 Contamination of seawater; 30
	Health/social; 75	Incidence of acute deaths; 40 Maximum number of deaths; 20 Incidence of delayed deaths; 20 Injured; 5 Evacuated; 15
Wastes; 25	Quantity; 50 Necessary period of embedding; 50	

energy depended on the significance of severe hazardous incidents (Hirschberg and Voss 1999).

What is noticeable from Table 7 is that the emissions are listed as a burden of normal operation (without quantifying their effects on health), while burdens resulting from severe hazardous incidents are placed with cases of death. The emissions figures, derived from an LCA, are thus regarded as an indicator of effects on health and the environment (Hirschberg and Voss 1999). Thus an aggregation of the aspect of "damage to health" remains open.

A further example of the application of a multicriteria analysis procedure as a result of the GaBE project is presented in Kröger et al. (2001). In this case the main criteria are taken from the "three pillars" model of sustainability. Accordingly, economic, ecological, and social aspects are either weighted equally or given particular emphasis as appropriate.

This shows, for instance, that, when economic or ecological aspects are emphasized, nuclear energy averages relatively well – better than natural gas, for example – but that it is poorly rated when social factors are brought to the fore. Admittedly, the health and environmental consequences of normal operation are not compiled with those of hazardous incidents and accidents here either. The assessment prospects of a future nuclear energy, for which the consequences of a possible accident and the embedding period of wastes were minimized, were also tested. Under these circumstances and with an equal weighting of sustainability criteria, nuclear fission was placed as the most favorable source of energy.

Conclusion

The works of the Poly project and, in particular, the GaBE project show that substantial and effective instruments for the comparative assessment of energy supply systems have been developed in recent years. LCA has formed the basis of this. The risks of energy supply systems – that is, the environmental health-related risks – form an aspect of the comparison. The well-informed choice between various energy supply options is an important goal of comparative assessments. Some aspects of the instruments, however, have yet to be improved and ensured. This refers to data about hazardous incidents – if this stems from retrospective investigations – whose case-specific applicability and significance are to be carefully examined. Regarding hazardous incident data from prognostic methods (such as probabilistic risk analyses), it should be transparent how these were calculated; standardized and quality-controlled procedures would be optimal here.

Particular difficulties lie in the aggregation of different facets of the findings of the analysis of a system into a useful overall evaluation. This is also the case for an integrated view of the operational and incident-related environmental health risks of energy supply systems. Generally, the aggregation must also integrate societal decisions about values. Thus approaches from the discussion about sustainability can be used, in the sense that sustainability, seen in broad consensus in its three dimensions (economic, ecological, and social) may present a framework for evaluations.

Table 8 gives an overview of the procedure.

Table 8 Procedure in the GaBE project.

Objective	A comparison of alternative technological methods; "public health", "public wealth"
Spheres of risk	Risks arising from facilities for the generation of electrical energy
Timeframes	In principle, unlimited
Geographic frame	In principle, unlimited
Attributes	Quantities of selected pollutants, land use, quantities of waste, and period of embedding, injuries and deaths through accidents
Aggregation/metric	Either financial assessment or a separate weighting of the attributes
Overall evaluation	Implicitly multiattribute (financial valuation) or explicitly multiattribute
Uncertainties	Unclear, in so far as examined. Bandwidths were given for indicators
Presentation of results	Table with values of attributes

3.2.6
Classification of Carcinogenic Airborne Pollutants for the German TA Air Novella

This project is presented here as one of the few cases of applied risk comparison in Germany (Schneider et al. 2001; Schuhmacher-Wolz et al. 2002).

The aim of the project is the classification of carcinogenic substances in one of the three classes of the TA Air Novella (Table 9) according to the strength of their effect, since on the grounds of commensurability the emission of particularly potent carcinogens should obey strict limits.

The evaluation is performed entirely on the basis of the strength of effect, i.e., no assessment is made of actual exposure or actual risk. To classify a pollutant, the unit risk value is consulted provided it is "very adequate" or "adequate"; in other cases the classification is made on the basis of available information which allows only a rough estimation of the carcinogenic potency in the range of high doses.

What is to be emphasized here is that the certainty of the assessment is brought into the evaluation along with the size of the risk: available unit risk values were evaluated for their quality (assessment certainty) and classified as "very adequate", "adequate", or "not adequate". A detailed catalog of criteria was consulted for this purpose (Schneider 2000).

Table 9 Procedure in the classification of carcinogenic airborne pollutants for the TA Air Novella.

Objective	Foundation for setting threshold limits for carcinogenic toxins
Spheres of risk	Cancer risk
Timeframes	Length of exposure, length of life of exposed individual
Geographic frame	Played no role, since concentration of pollutant was considered
Attributes	Cancer
Aggregation/metric	Unit risk
Overall evaluation	According to unit risk; classification into one of three classes
Uncertainties	Specification and examination of uncertainty bandwidths of unit risk values
Presentation of results	Table with unit risk values and classifications

3.2.7
Summary

One may conclude from the six case studies discussed that risk comparisons may be differentiated with regard to their goals, formation of risk categories, timeframes, metrics of comparison, and constraints.

Goals. Typical goals of risk comparisons are the establishment of priorities in environmental health policy, the choice between technological or administrative alternatives, and the setting of threshold values.

Formation of categories. A typical example for the problems of category formation is the US EPA Unfinished Business project. Here the comparative categories are oriented on the lines of the internal structure of the EPA, which in turn must be oriented on its role as laid down in law. It is not negligible how these are categorized, because the broader a category is, the greater is the chance that it will take a higher place in the ranking. The categories defined in the Unfinished Business project have the further disadvantage that emission sources, effect pathways, and effects are all used as categories, so that some risks are counted twice.

Timeframes. The averaging over time of the consequences of rare accidents that may cause substantial damage, in order that they can be compiled together with operational risks, is problematic (see ExternE project). We propose that such contributions to risk be separately compared.[39]

Metric of comparison. In projects with the aims of "establish priorities in environmental health policy" and "technical or administrative alternatives" are risks that distinguish, for example, between the types of harm or damage, timing of damages (the question of discounting), and groups of people of varying sensitivity. For these contributions to be aggregated and compared, they must be converted to a common metric. Because in so doing valuations that are basically subjective must of necessity be made, the metric used is often criticized. The metric therefore requires legitimization.

In the project with the aim "setting threshold values", a criterion of comparison ("risk of cancer") is at hand for which an established metric (unit risk) is available.

Constraints. The risk under investigation is dependent on many constraints, such as the spatial distribution of the subjects of protection (population, ecosystems, material assets) in the vicinity of the source of risk, previously existing burdens from other sources of risk (this is particularly relevant in the case of pollutants with nonlinear dose–effect relationships), interactions which lead to the decomposition or formation of pollutants, such as that established in ExternE between SO_2, NO_x, and NH_3.

It follows from this that statements of risk are only ever valid for the total scenario set out in the analysis or statistical survey.

In Chapter 4 the issues and problems discussed here are more precisely investigated on the basis of empirical risk and decision research.

---

[39] See Chapters 5 and 6 for details.

4
The Empirical Foundations of CRA

4.1
A Starting Point for Risk Comparisons

The technical prerequisites for a risk comparison are simple: the risks being compared must possess characteristics of at least one common attribute that will permit a comparison to be made.[40] A further requirement is that the variation of the characteristics may not be so great as to render an unambiguous order of risks (less than, greater than, equal to) unfeasible. For instance, the risks arising from genetically modified organisms can be compared on an unequivocal scale with those of meteorite strikes if deaths occurring in both cases are considered and can be accurately estimated. In addition, the relevant data must also be available.

Which comparisons are, however, meaningful is a separate issue. The question of which risks can be compared must be rephrased as: why should the chosen risks be compared? The clarification of these prerequisites, relevant both to decision-making and operations, is of central significance and is discussed below.

It is widely considered indisputable that the risks of alternatives are to be compared. Such comparisons have a long tradition in the field of technological assessment. It is also common to evaluate various therapies through a comparison of their associated risks. In technology, for example, locations or technological options for the power industry (Keeney 1980; see also Kunreuther and Linnerooth 1983), in waste management (Powell 1996), as well as the options for the transport of hazardous goods (road versus rail) have been compared. Comparisons between various technological options (e.g., nuclear energy versus fossil fuels), however, encounter the problem that risks may be compared which on the one hand are a combination of very low frequencies of incidence with very great damage and,

---

[40] In comparisons a minimum of two objects are placed in mental connection so as to identify the similarities, differences, or shared properties based on specific aspects of their characteristics or relationships. On this basis, for instance, orders may be formed, i.e., groupings or rankings or series according to one or more contextual or formal characteristics (cf. *Fachlexikon der Psychologie*, Deutsch Verlag, 1995 [Translated from German]).

Comparative Risk Assessment. Holger Schütz, Peter M. Wiedemann,
Wilfried Hennings, Johannes Mertens, and Martin Clauberg
Copyright © 2006 WILEY-VCH Verlag GmbH & Co. KGaA, Weinheim
ISBN 3-527-31667-1

on the other hand, a combination of relatively high frequencies of incidence with less damage. Put another way, the issue is how to weigh potential catastrophes.[41]

Entirely disparate risks may also be compared. Thus highway construction, urbanization, environmental pollution, and the application of biotechnology to crop production can all be compared by the criterion of "loss of biodiversity" so as to set priorities or to identify causes for concern.

Furthermore, the risks of medical examinations using ionizing radiation can be compared with those arising from exposure to radon. The aim here could be to establish the significance of the risks.

The examples show that it depends on clarifying the operational context and defining the focus of the risk comparison in this context.

Basically, one can distinguish three operational contexts, which are laid out in Table 10.

Table 10 Contexts of risk comparisons.

Focus	Operational context	Aim
Communication problem	Understanding of risk	Relaying the significance of risks
Prioritization problem	Risk assessment	Ranking/categorization
Decision-making problem	Risk management	Best option/variations/reason for action

These differing purposes of comparisons are considered in more detail below. Here we will concentrate on two areas: comparison of risks as a means of risk communication, and problems of CRA procedure in establishing priorities. Moreover, the organizational problems of CRA and the communicative problems of relaying findings are discussed.

4.2
Risk Comparisons as a Means of Risk Communication

Risk comparison is taken here to mean the comparison of a known with an unknown risk. Such comparisons serve primarily to ensure understanding. As mentioned above, comparisons require that the objects being compared have at least one attribute in common by which they can be compared.

Comparisons have always played a special role in risk communication. Two or more risks are contrasted to demonstrate their nature, magnitude, and relative significance. At its core, this was to arrive at an appropriate evaluation of the "riskiness" of a risk and to minimize any harm caused by consternation or alarm. However, it quickly became clear that these expectations were not to be realized very easily.

[41] It would thus be consistent to require that a hazardous incident analysis for fossil fuels should also be conducted on a worst case basis.

The first methodically deliberated suggestions for risk comparisons stem from the year 1988. In that year the US Chemical Manufacturers Association published *Risk Communication, Risk Statistics and Risk Comparisons: A Manual for Plant Managers* (Covello et al. 1988), which included a chapter on the requirements of successful risk communication.

It is the authors' opinion that comparisons help to put risks in perspective. They also assist people to a better understanding of the significance of a risk. However, it is shown that risk comparisons can be delicate, since they can be perceived as an attempt to make clear to a concerned public that they are worrying about the wrong risks. It is fairly obvious that such lecturing can be badly received. It was, therefore, the intention of Covello et al. (1988) to show which risk comparisons make sense and which do not.

They thus focus on the reassuring type of risk comparisons, i.e., those which show that a particular risk is, in comparison to another, of far less significance. In their opinion warning risk comparisons are somewhat simpler: they meet with fewer caveats. They order risk comparisons into five levels designed to express the acceptance they meet:[42] the lower the level, the less likely is the comparison to find acceptance (see Table 11).

[42] "It is possible to rank different kinds of risk comparisons in terms of their acceptability to people in the community" (Covello et al. 1988, p. 17).

Table 11 Levels of acceptance of risk comparisons. (After Covello et al. (1988).)

Risk comparisons

Level 1 (highest acceptance):
  Comparisons of the same risks at different points in time
  Comparisons with a standard (threshold)

Level 2:
  Comparison of the risks of taking an action with those of failing to take the action
  Comparison of risks arising from various solutions to a problem
  Comparison with the same risk in a different location

Level 3:
  Comparison of average risk with an excessive risk
  Comparison of the risk of an adverse effect from one source with that from all other sources of the same adverse effect

Level 4:
  Comparison of risks with costs, i.e., risk–cost relationship
  Comparison of risk with benefit
  Comparison of work-related with environmentally related risks
  Comparison with other risks from the same source
  Comparison with other causes of the same harm or damage

Level 5 (lowest acceptance):
  Comparison of risks between which there is no connection

The model on which Table 11 is based is relatively simple: risk comparisons ought to be more acceptable the more plausible and clear the comparison of common aspects appears to be, in order to make a meaningful comparison of the two circumstances possible. The comparison of a risk with a standard, for instance, is held to be more appropriate than a comparison of a particular risk with the risk arising from another source that causes the same type of harm or damage. In particular it would follow from the model of Covello et al. that CRAs of the types found in levels 4 or 5 are bound to encounter great difficulties. However, this is not the case. A survey of the acceptance of CRA procedures from the USA (Charlton Research Company 2001), in which 800 people were asked whether they considered CRA to be a good idea, found that 39% of those questioned considered CRA to be a good idea (strong approval) and another 41% gave weak approval to the idea. Only 11% rejected the idea.

However, the model of Covello et al. has other problems, as the following example will illustrate. Let us suppose that a resident has a radon survey conducted, which discovers a concentration of 800 Bq m^{-3}. The SSK (1994) has established a decontamination level for radon; decontamination is necessary at any concentration above 1000 Bq m^{-3}. The comparison thus would consist here of a contrast of measurement and action values. This "level one" comparison is, according to Covello et al., better than the following "level four" comparison, taken from a Swiss leaflet: "According to our current knowledge, radon is the most common cause of lung cancer after smoking". However, the latter informs in a convincing way about the magnitude of the risk.

In our opinion the example shows that the acceptance levels cannot simply be assigned on the basis of the similarity of the risks. It is also relevant how well known the reference risk is and whether the comparison suggests courses of action.

Nonetheless, the model proposed by Covello et al. has created a surprisingly strong resonance and even today is to be found in overviews of, and recommendations about, risk communication (cf. WHO 2002). These, however, overlook the fact that this acceptance ranking is not supported by any empirical experiments. In any case, it should be noted that four different aspects of risk comparisons are to be distinguished:
- acceptance of the risk comparison;
- the contribution that the comparison provides to the understanding of risk;
- the effect of comparisons on risk perception;
- the influence of risk comparisons on the acceptance of a risk.

These aspects can be linked to a cascade model: firstly, the comparison has to be accepted. Only then is it possible for its impacts to develop. A first cognitive effect can lie in the improvement of understanding. Various possibilities may be envisioned: magnitudes become more understandable, the significance of a risk can be better estimated, or the extent of the uncertainty of a risk becomes understandable. Only when these cognitive preconditions are satisfied can an influence on risk perception be anticipated. A change in the perception of risk can, though it need not, lead to a change in the acceptance of risk.

An early study, which examined the effects of comparisons related to the risks associated with radon, was conducted by Weinstein et al. (1989). They were able to demonstrate that comparisons improve the understanding of risk.

A further study was conducted by Roth et al. (1990). They constructed risk communication scenarios in which participants were asked to advise a manager about the choice of risk comparisons for a presentation. A total of 14 various risk comparisons were given, which the participants had to assess on seven scales (clarity, aid to understanding, usefulness of information, over- or undervaluation of the risk, reassurance, formation of trust, and a "should be used" scale). The 14 risk comparisons were taken from the five levels of the categorization of Covello et al.

All the risk comparisons were evaluated as of similar value on the "should be used" scale (scoring between 1.81 and 2.92 on a seven-point scale on which the best was represented as 1). An analysis of correlation between the scores on the seven scales and the risk comparison levels of Covello et al. found consistently negative results except for the "formation of trust" scale. Here there was practically no correlation ($r = 0.1$). Clearly the acceptance levels do not play the part they are postulated as playing.

Roth et al. (1990) reach the conclusion that the acceptance of a risk comparison does not depend on the similarity of the risks compared. Experiments by Johnson 2003a support this assessment. This study also found no correlations between evaluations using the seven scales (clarity, aid to understanding, usefulness of information, over- or underrating of the risk, reassurance, formation of trust, and a "should be used" scale) and the risk comparison levels. In addition, Johnson also constructed a negative scenario. The participants were informed that the risk manager would meet with mistrust and rejection. No relationships were discovered in this context either. The model of Covello et al. (1988) thus rests on a somewhat shaky empirical foundation.

Slovic et al. (1990) report a further experiment in which three variants were given. Under condition A the participants were given a simple risk scenario (asbestos exposure in a school, control condition), under condition B risk comparisons, and finally under condition C in addition to the risk comparison a critical commentary which placed the risk comparisons in question. This showed an attenuated risk perception (as a consequence of the risk comparison) under condition B, which did not occur under condition C. The experiment, though designed to defend the approach of Covello et al., does not support the risk comparison levels. It rather makes clear how complex are the conditions upon which the effects of risk comparisons depend.[43] Clearly contextual factors play a major role. This experiment was also repeated by Johnson (2002). In an extensive study he tested whether various conditions of the information specification exhibited an influence on the impact of risk comparisons. In so doing, he could replicate the findings of Slovic et al. In addition his results demonstrate that an inoculation (i.e., anticipating criticism of risk comparisons) limited the effect of such criticism.

---

[43] Unfortunately Slovic et al. (1990) conducted no significance tests. This limits the power of their statements.

A third experiment was performed by Freudenburg and Rursch (1994). The authors provided an imaginary risk scenario (the construction of a hazardous waste incinerator) and invited the subjects to put themselves in the position of local residents. First, the subjects' attitude to the construction of the incinerator was asked (63% were opposed). Then they were given a statement (an independent study, commissioned by the industry, finds a risk of 1:1,000,000). Some 36% of the participants then said that they would sooner support the construction of the incinerator (no change: 54%; less support: 8.5%). Following this a risk comparison was presented (1:1,000,000 is a smaller risk than smoking a few dozen cigarettes). Here 62.1% said that this comparison did not lead them to alter their views, 18.2% were then ready to support the construction of the incinerator, and 19.3% were less prepared to support it.

Johnson and Chess (2003) examined the effect of comparisons with a standard on the risk evaluation. The first point to note is that a third to a half of respondents expressed a negative attitude to threshold values. They did not, for instance, believe that threshold values really protected health, or they assumed that greater attention was paid to costs for industry in the establishment of such threshold values than to the protection of health. Various exposure scenarios were presented in the experiment, which utilized the threshold value differently (95 or 50% of the threshold value). In every treatment, the participants were unsettled and assumed that they faced a serious risk. Risk comparisons, which make use of threshold values, do not, it appears, have a reassuring impact.

The experiments described here demonstrate that the acceptance of risk comparisons is relatively high and does not follow the order laid out by Covello et al. (1988). Participants in the experiments of Roth et al. (1990) and Johnson (2002) assessed the acceptance of comparisons, with one exception, as between 2 and 3 on a 7-point scale (whose best value was 1). The contribution of risk comparisons to the understanding of risk is – see the data of Roth et al. (1990) and Johnson (2002) – also good (values between 2 and 3 on a 7-point scale whose best value was 1), independent of situational context (i.e., resistance or a neutral–positive climate). However, the impact of risk comparisons on risk perception is somewhat marginal. Apart from a few exceptions, they have only nonsignificant effects. This explains to some extent the negative impact, or the absence of positive impact (presumed and as yet empirically unsupported) on the evaluations of the acceptance of risks. Changes in acceptance may occur in two ways: either through a change in the perception of a risk or through a change in the standard of acceptance. The effects of a risk comparison thus in no way depend solely on the similarity of risks. The perceived purpose of the comparison and the consistence or inconsistence of the information provided about the magnitude and significance of a risk apparently plays a far greater role.

In summary, it appears that risk comparisons
- are accepted
- have a clear impact on the understanding of risk
- have hardly any impact on risk perception
- evidently have hardly any impact on the acceptance of risk.

From this it follows that there is no scientific evidence that argues against the performance of a CRA. Although the impact of risk comparisons on risk perception is slight, it must be remembered that an integrated CRA also includes the participation of citizens: they are not simply the consumers of risk comparisons. The findings also suggest that the acceptance of CRA procedures, the contribution of CRA to the understanding of the magnitude of risks, the influence of CRA on the perception of risk, and the acceptance of the findings of CRA should be distinguished and regarded as areas of duty in the development and performance of a CRA.

4.3
Procedural Challenges

Florig et al. (2001) provide evidence showing that a large number of CRAs performed so far have used inadequate standards, and are thus not satisfactorily valid. They propose a stronger consideration of the results of decision-making research, the insights of research into the perception and communication of risk, and psychological inquiries into human information processing.

We would like to provide here an overview of these challenges. Central aspects include the determination of underlying concepts, the formation of risk categories, the choice of attributes, the assessment of options, and the performance of comparisons. The errors and pitfalls of a participative comparative risk evaluation, following empirical findings, should be shown here. This part is especially important, because we believe that not only experts but also representatives of societal groups and lay people should participate in CRAs.

4.3.1
Framework: What is to be Kept in Mind when Determining the System Boundaries for CRA?

Risk comparisons imply delimitations: geographical and chronological limits, as well as limitations of content, are necessary. Geographically, regions are to be defined (e.g., state, region, district) for which risk comparisons are to be made. The same is true of chronological periods. It makes a difference whether currently existing risks (retrospective view) or future risks (prospective view) are in question.

The framework for a risk comparison is not naturally given, but is – depending upon the goal – constructed or selected. Thus different limits are drawn, focused on various classes of risk and suggested by different reference points and various categories of measurement.[44] Thus it is conceivable that one may proceed from rather cautious, and another from somewhat more optimistic, assumptions. Table 12 shows an example of this.

[44] One instance of the susceptibility of measurement categories to framing effects would be cases in which risks are compared on the basis of relative risks and, thus, incidents that are actually significant are ignored.

Table 12 Example of the impact of framework.

	Frame A	Frame B
Attitude	Cautious	Optimistic
Cause	Focus on multiple stress factors	Focus on a single stress factor
Exposure	Inclusion of various exposure paths	Inclusion of assumed main exposure path
Consequences	Analysis of second-order consequences (consequences of consequences)	Analysis of first-order consequences
Evidence	Focus only on positive results	Focus on overall scientific picture, i.e., positive and negative results

The framework has a considerable influence on the following steps, notably on the assessing and comparing, as psychological decision-making research (which in this context uses the term framing effects) shows (see Russo and Schoemaker 1990; Schoemaker and Russo 2001).

Schoemaker and Russo (2001) emphasize the framing traps in this connection. These include: (a) the illusion of complete problem representation, (b) overconfidence in existing knowledge and the consequent dominance of one's own frame, and (c) the tendency to consider only such information that supports one's own frame.

In connection with the first framing trap, the investigation of Fischhoff et al. (1978a) is instructive. They demonstrate that people overestimate the completeness of a problem's representation; they overlook omissions. It is easy to imagine that that this error can also occur in the collecting of risks for a comparative assessment. Whether a representation of a risk sphere, for instance all the risks for an ecosystem or all the health risks in a region, is complete, or even adequate, can be falsely evaluated. Overconfidence plays a role here: overconfidence is the tendency to be too sure that one's own assumptions and appraisals are right. It has been shown in a series of experiments that this effect is also to be reckoned with in experts when they make appraisals (cf. Bazerman 1990). The tendency to prefer confirmatory information is also to be seen in experts (Einhorn and Hogarth 1978).

A significant consequence remains; with the collection of the problem list for comparative risk evaluation, important courses are set. Great diligence must therefore be brought to an adequate representation of the problem. It should be attempted, from various approaches with different central questions, to arrive at a more complete and structured list of sources of risk, selected by consistent criteria. It is advisable to proceed in a theory-led and issue-related fashion and create homogeneous classes, which emanate from a single link in the chain of risk. As a rule, these will be events (emissions) that may be assigned to a single source of risk (substance classes, e.g., emissions from traffic).

4.3.2
Risk Categories: What Influence Do They Have on Comparisons?

It is all but trivial to state that objects to be compared play a central role in comparisons. As ever, the devil lies in the detail.

A glance at the lists of risks of completed CRAs reveals that as a rule very heterogeneous risk categories are addressed. Besides environmental, health, and welfare problems one finds indoor air pollution, waste, lack of environmental awareness, global climate change, loss of inner city green spaces, and others. Technologies and other matters are also listed (e.g., waste incineration and AIDS). The US EPA Science Advisory Board (1990), in its critique of the Unfinished Business project, has already discussed the problems of measurement and comparison, which arise when such heterogeneous reference systems are used. This statement can be illustrated by two examples.

The risk ranking of the Environmental Strategies project for Metro Denver[45] is the first example. It is noticeable here that risk categories such as sources of risk (e.g., waste deposits), as contaminants (e.g., air pollutants), and also as resources to be protected (drinking water, surface water) are given. This leads to overlaps. Thus, for instance, a risk source such as a waste deposit (Superfund Site) can also have effects on drinking water. It is likewise possible for underground storage tanks to have an influence on the quality of ground water. Sources of error creep in through the reference system.

In risk ranking in Florida, in contrast, entirely different risk categories are applied (cf. Florida Center for Public Management 1995). Here the objects to be protected, such as indoor air quality, food quality, and scenic, historical, and cultural resources predominate. Risk drivers such as "patterns of development" and "use and management of public lands" are also found. In this way, inconsistencies appear; it is not clear, for instance, which risks influence food quality, or what effects the given risk drivers have.

These categorization problems will be addressed below from two points of view. One of these is the influence of the formation of cognitive orders; the other is the affective influences, which may result from the description and characterization of risk classes.

For a comparison objects, in our case risk categories, are necessary. Comparisons on the basis of single substances are not always necessary, since 10,000 substances or more may need to be compared. Therefore, risks have to be grouped. What is an appropriate grouping? This apparently simple question is not easy to answer.

Morgan et al. (2000) propose a classificatory scheme that is geared to the risk chain of Hohenemser et al. (1985). It is their opinion that the chain from risk-bearing activities to their final consequences for affected people can be used to select a set of risks for CRA. For instance, one may commence by gathering together the

---

[45] *Comparison of Human Health Risks and Economic Damages from Major Sources of Environmental Pollution in Denver*, Industrial Economics, Inc., 1998.

activities leading to air pollution in a region: exhausts from traffic (cars and trucks), coal-fired power stations, chemical plants, etc. These can be further specified according to the pollutant emitted, and they can then be grouped depending on the endpoint-specific pollutant groups. A similar proposal has been suggested by the US EPA (1993a).

In conclusion, the categories of such groupings should
- be logically consistent (exhaustive, not overlapping, and sufficiently homogeneous that they allow evaluation by selected attributes);
- have cognitive biases reduced to a minimum (easing understanding and minimizing framing effects) and the limits of information processing considered (e.g., a manageable number of categories);
- consider regulative concerns (availability of data, competence and feasibility in management options);
- ensure fairness (considering the concerns of all stakeholders and the general public).

Such suggestions may be refined by reference to empirical investigations into risk perception. For that purpose several experiments are presented below.

In one study of the perception of the acuteness of environmental problems, Balderjahn and Wiedemann (1999), proceeding from categorizations of environmental problems, conducted a correspondence analysis. This analysis demonstrated that environmental problems with a high degree of abstraction, such as air and water pollution, were seen as requiring more urgent attention than those with a lesser degree of abstraction. Environmental problems with a low degree of abstraction (pollution of the Rhine, extinction of storks, etc.) were classified as "uncertain regarding acuteness". This leads to a conclusion for public risk appraisal (Karger and Wiedemann 1998): abstract problems are more likely to be considered as a "great danger to people" than problems with a lower degree of abstraction.

Besides such problems of cognitive grouping, there are also affective aspects of categorization, which can influence risk perception. The naming of key terms as, for instance, genetic, atomic, or chemical, can trigger a series of negative associations. Such stigmatization can then have an impact on the assessment of risks (cf. Gregory et al. 1995; Murphy 2001; Wilson and Crandon 1998). To give one example, Lindell and Earle (1980) have shown that the same risk estimate, presented once anonymously and named once as "nuclear power", leads to different evaluations. A survey in the UK (1999) came to a similar finding. It shows that the substitution of the term "organic farming" with the term "carbon-based biotechnology" in interviews leads to calls for stricter regulation. We have been able to replicate this result (Wiedemann and Brüggemann 2001).

Our own experiments indicate that such emotional framing effects are not only found on the level of words, but can also occur through an embedding in various stories (Wiedemann et al. 2003). It could be demonstrated in a series of experiments that the presentation of an identical risk in stories of various emotional tones leads to different assessments (see Figure 11).

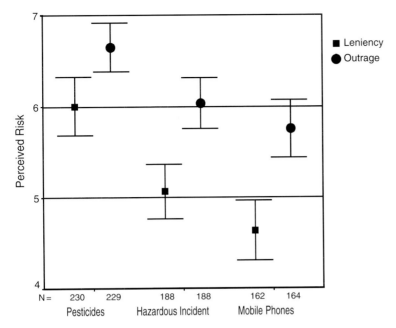

Figure 11
The effect of risk stories on risk perception.

In such stories the people involved (e.g., the generators of risk and the victims), their actions, and their motives and reactions were characterized. Thus the emotionalizations[46] (e.g., leniency versus outrage) can be conveyed, which influence risk perception. These effects are stable for various risk scenarios (risks due to hazardous incidents, pesticides, cellular phone networks). It appears that a halo effect occurs.[47] The negative presentation of the generators of risk leads to emotions, which in turn affect the assessment of risk.

This has the consequence that in risk comparisons the ranking does not follow the risk attributes but is influenced by the properties of the context (which could perhaps only be imagined). We have shown in earlier experiments (Schütz et al. 2000) that such factors added by the context properties of interviews are inevitable.

The available evidence clearly indicates that the categorization and labeling of risks exercises an influence over risk assessments and therefore over risk comparisons. It is advisable that labeling is as neutral as possible and that stigmatizing terms are avoided. Nonetheless, the influence of affective dimensions on risk comparisons can never be entirely excluded.

[46] For the experiment, stories were constructed which described a risk. Different facets of specific elements of the stories were presented. For instance, the size of businesses involved was varied, as were the exposure to risk and the residents. These stories were then given to participants in the experiment for their appraisal.

[47] Halo effect: describes the tendency of assessments of notable characteristics to affect the assessment of other characteristics.

4.3.3
Attributes: What Influence Do They Have on Comparisons?

Similar to the creation of risk categories, there are a multitude of possibilities with regard to risk attributes. The CRAs conducted thus differ considerably from each other. Thus, along with ecological and health outcomes are attributes such as peace of mind, fairness, and effects on future generations (cf. the scorecard approach of the Environmental Defense Fund).

From the point of view of decision-making theory, at least three requirements need to be fulfilled when choosing attributes: (a) the number of attributes should be manageable; only a limited number of attributes are ever referred to in risk comparisons; (b) the attributes must be independent of each other in their assessment (see Section 6.2.3); and (c) the chosen attributes should be relevant to the assessors.

A particular challenge for a participative CRA is the requirement of relevance, since the findings on risk perception suggest that different attributes may be relevant for lay people and experts. From this it follows that, besides scientific criteria, lay criteria should also be applied. It is debatable whether this applies and whether this requirement for the inclusion of lay attributes can be coherently met (Sjöberg 1996; Schütz and Wiedemann 2003).

We have therefore pursued this question and reanalyzed data from earlier experiments (Wiedemann and Kresser 1997; Wiedemann and Balderjahn 1998). This showed that the attributes that are considered relevant from the point of view of psychometric risk research, and which, for instance, Morgan et al. (2001) use in their approach to CRA (potential for catastrophe, controllability, uncertainty of scientific knowledge, and number of individuals affected), are not those that lay people use when they are given the freedom to choose attributes for themselves.

In a further experiment that we conducted, a list of environmental problems was given to participants to be classified into three categories of urgency (urgent, less urgent, do not know).[48] From these classifications the attributes by which the participants had categorized were ascertained. Attributes such as problem pattern, characteristics of damage, and characteristics of problem management predominated.[49]

An experiment by Morgan et al. (2001) corroborates our opinion that what are termed qualitative factors play no particular role in risk ranking. When risk attributes are evaluated according to their importance, classic attributes such as mortality and morbidity play a greater part than controllability of the risk, uncertainty, quality of scientific knowledge, and latencies.

---

[48] The list comprises 30 environmental problems and includes abstract environmental problems as well as concrete damage (e.g., damage to oak trees, pollution of the North Sea).

[49] In the sphere of problem pattern it is largely the extent (global–local) and the magnitude (general–specific) of environmental problems which play a role, as well as knowledge of the problem. Beyond that, the cause is significant. When participants name consequences, they focus most strongly on consequences for people, then on consequences for the environment. If problem management is picked out as a central theme, stress falls on controllability and the solutions to an environmental problem.

> **Box 1: Ranking of evaluative criteria according to importance**
>
> 1. Consequences for health and lives of people
> 2. Irreversibility of possible environmental damage
> 3. Nature and strength of possible environmental damage
> 4. Certainty of controllability if this risk
> 5. Probability of damage occuring
> 6. Level of awareness of risk
> 7. Personal impact in the event of damage
> 8. Economic necessity of the acceptance of the risk
> 9. Degree of public and social conflict (media activity) in the event of damage
> 10. Nature and extent of economic risks

In a further experiment conducted by Balderjahn and Wiedemann (1999), we specified attributes for risk assessment. Experts, lay people, managers, and administrators were interviewed. The interviewees' task was, for each of the sources of risk, to say how significant or insignificant the attributes were for the assessment of the risk in question.

For all the sources of risk, a summary, ranked by importance, of evaluative criteria is given in Box 1.

The differences between the estimations of the significance of single criteria are slight and in no case significant. The significance of evaluative criteria also varied little between sources of risk.

We have, in experiments based on the works of Balderjahn and Wiedemann (1999), likewise ascertained what significance different characteristics of attributes have for various groups (experts, lay people, managers, and administrators). The participants had to evaluate a hypothetical environmental risk by reference to a series of evaluative criteria, which were presented in various distinguishing characteristics. Damage to health and to the environment, the probability of damage, the profile the problem has among the public, and the increase in employment were all varied. The participants' task was to say how acceptable the risks associated with the possible combinations of characteristics were.

The significance of the attributes hardly differs between the separate groups. The probability of damage was the most important decision criterion for all groups. While this criterion accounted for about one third of the decision for administrators, experts, and lay people, the equivalent figure for managers was only a quarter. The second most important attribute is the consequences for people (~25%), followed by increase in employment and consequences for the environment. Only for experts the criterion of consequences for the environment was more important than increase in employment. The criterion of public debate was relatively unimportant; only managers considered this criterion somewhat more strongly.

Lion et al. (2002) have investigated which information relating to unknown risks lay people prefer. The following risks were applied: genetically modified foodstuffs, radon concentration in dwellings, a new blood clotting agent, dioxin emissions from waste incinerators, and electromagnetic fields. It showed that lay people are primarily interested in the following issues: "How is one exposed?", "What are the consequences of the risk?", "Wherein lies the risk?", and "How great is the probability?"

These empirical findings demonstrate how difficult it is to answer the question of which risk attributes are "naturally" preferred. Even minor changes in question format or experiment design suggest various different risk attributes.

An inclusion of the qualitative risk factors which psychometric risk research would recommend (potential for catastrophe, awfulness, profile of the risk, knowledge about the risk in the sciences, voluntariness, etc.), as suggested by Fischhoff et al. (1984), among others, and tested by Morgan et al. (2000, 2001), among others, is not advisable. The available research findings argue against this.

Attributes must be derived from aims (see Keeney 1992). In this the fundamental subjects of protection of human health, intact ecosystems, and quality of life must be considered. On the level of attributes, the probability of occurrence – contrary to the assumptions of psychometric risk research – also plays an important role for lay people. It is also, furthermore, advisable to consider the degree of the problem (the seriousness of effects), the extent of the problem (how many people are affected, how widespread are the environmental problems), and perhaps the quality of the problem (the reversibility or irreversibility of the problem).

4.3.4
Assessment of the Attributes: What Influence Does the Measure of a Risk Have on Comparisons?[50]

Different measures of risk may be applied to attributes to characterize risks. For instance, for a technological risk one can give the number of plants in existence that present such a risk. Further, the probability per lifetime of each plant may be considered along with the pollutant released by a hazardous incident. Risks may be characterized by reference to particulars other than pollutants. Specifics of quantities, concentrations, or intensities, or the extent (what area is affected by pollutant) can be characterized. Finally, the persistence – how long the pollutant pollutes – is characterized in measures of duration (e.g., half-life).

Exposure can be characterized by, for instance, how many people are affected. Furthermore, accumulation over time can also be presented: in what quantity does the pollutant concentrate in bodies or the environment over a particular time span? The adverse effects can also be variously described. Specifics of acute or chronic effects, of frequency of sickness and of the nature of possible harm to health, of the effects on future generations, or of damage to the environment can all be used. Naturally, the choice of measure will depend on whether it meaningfully characterizes the risk, and on whether data on the measure are available.

[50] This section was derived in collaboration with Dr. Andrea Thalmann, whom we thank for her supporting efforts.

Experimental investigations demonstrate that the choice of measure influences the characterization and risk perception (cf. Femers 1993; Gray 1996). An example from Crouch and Wilson (1982) illustrates this. If one wishes to determine whether US coal production became safer in the period 1950 to 1970, one can use the number of accidental deaths per million tons of coal extracted. This shows that coal production became safer over the period in question. However, one can also examine the number of deaths per thousand workers. In this case, it appears that coal production over these 20 years became less safe (see Figure 12).

Another instance is the oft-quoted comparison of the risk of flying with that of driving. If one calculates from the number of deaths due to aircraft or traffic accidents per kilometer (or mile) traveled, one finds a significantly higher risk for driving: 13 deaths per 10^9 miles driven, as opposed to 0.6 deaths per 10^9 miles flown. However, if one considers only journeys that are shorter than 600 miles (this is, at least in the USA, a great proportion of flights and an even greater proportion of car journeys), one arrives at a lower risk for driving.[51]

Table 13 summarizes the effects of various risk measures on risk perception. For example, Stone et al. (1994) found that risk information that was delivered relatively (e.g., a doubling of the previous risk) exercised a stronger influence over judgment than a description in the form of an incidence rate. This applies above all to very low probabilities. Thus extremely low incidence rates are perceived as "practically nil", while this was not the case of statements in the form of a relative risk. According to the experiments of Stone et al. (1994), people are ready to pay more for their safety when the risk reduction is presented relatively, compared with a presentation using the aid of an incidence rate. The findings of Magat et al. (1987) tend in the same direction.

[51] The example comes from Evans et al. (1990). The data relate to the time period 1978 to 1987. The authors criticize the application of mean values to accident statistics, which, particularly for car drivers, masks the great variability of frequencies of accidents for people and age groups. In fact such calculations cannot be made without a whole series of further assumptions which are extensively discussed and their effects on risk comparison presented by the authors.

Table 13 The effects of characterizations of risk on risk perception.

Characterization of risk	Effect on risk perception
Probability	
Relative risk	→ tends to strengthen
Base rate (with small risks)	→ tends to weaken
Positive verbal statement of probability	→ tends to strengthen
Negative verbal statement of probability	→ tends to weaken
Approach to uncertainty	
Presentation of confidence intervals for risk estimates/description of uncertainties	→ tends to strengthen
Consequences	
Seriousness/degree of severity	→ tends to strengthen

92 | *4 The Empirical Foundations of CRA*

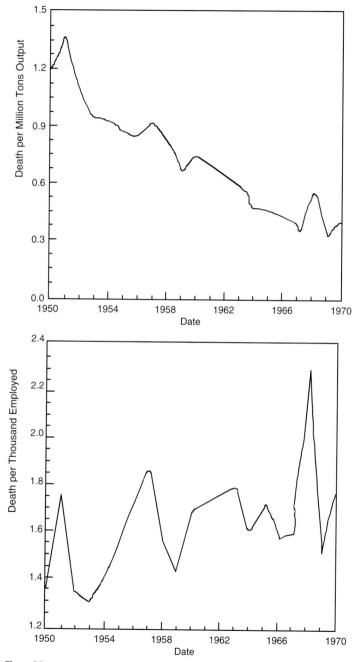

Figure 12
The influence of selected indicators on the presentation of risks.
(From Crouch and Wilson (1982, p. 99).)

Halpern et al. (1989) also examine the influence of various presentations of information on the appraisal of risk. The authors assert that the representation of a probability as a relative risk raises the evaluation of that risk, compared to a representation of the same probability referring to the frequency of deaths. Halpern and colleagues say that the explanation of this discrepancy is that those interviewees who knew the frequency or the absolute numbers of deaths appraised the risk differently due to knowledge of its basic rate.

A series of studies have identified the basic rate (e.g., Wallsten et al. 1986a; Weber and Hilton 1990; Fischer and Jungermann 1996) as a central variable influencing the interpretation of verbal characterizations of probability. For instance, Weber and Hilton (1990) showed that the interpretation of verbal statements of probability was influenced by knowledge of the basic rate. This was also the case in the findings of Wallsten et al. (1986a), who were likewise able to identify the perceived basic rate as a factor influencing the interpretation of statements of probability.

Verbal statements of probability were often thought to be too vague (cf. Budescu and Wallsten 1985, 1995; Fillenbaum et al. 1991; Wallsten et al. 1986b; Zimmer 1983).

Contextual effects on the numerical interpretation of verbal statements of probability such as "few", "little", "many", or with frequency, e.g., "rarely", have been found in a series of studies (Beyth-Marom 1982; Brun and Teigen 1988; Budescu and Wallsten 1985; Fillenbaum et al. 1991; Gonzales and Frenck-Mestre 1993; Hamm 1991; Teigen 1988; Weber and Hilton 1990). Thus, for instance, the phrase "I often visited the cinema" is assigned a different numerical value from "I often visited the USA" (Newstead and Collis 1987). Likewise, the meanings of "seldom", "occasionally", and "frequently" were interpreted differently in various contexts.

Verbal expressions are less neutral than numerical statements and therefore can influence decisions in subtle ways (cf. Budescu and Wallsten 1995; Moxey and Sanford 1993; Champaud and Bassano 1987). Thus there are expressions that accentuate the existence of an occurrence, such as "probably" or "possibly", while others emphasize their nonexistence, such as "doubtful" (cf. Teigen and Brun 1999, 2000).

In connection with the use of verbal and numerical statements of probability, Erev and Cohen (1990) as well as Wallsten et al. (1993) describe the "communication mode preference (CMP) paradox". This states that most people prefer to receive information about the probability of occurrences in the form of numerical statements – but prefer to use verbal statements of probabilities themselves.

The study of Erev and Cohen (1990) demonstrates that people prefer verbal terms to assess imprecise probabilities regarding the incidence of occurrences. The preference is perhaps based on the assumed congruence between the precision of an expression and the underlying uncertainty (cf. Teigen and Brun 1999). Furthermore one cannot forget the great variability between individuals in the meanings assigned to words expressing probabilities, which is to be found among both lay people (Thalmann 2005) and experts (cf. Jablonowski 1994).

In our own experiments (Thalmann 2005) we investigate the meaning of statements of probability among lay people. These deal with the verbal indicators of the German Radiation Protection Commission (SSK 2001), which are used for the characterization of existing evidence concerning the health effects of electromagnetic fields.

It turns out that the participants evaluate "hint", "suspicion", and "proof" differently from that foreseen by the SSK.[52] Rather than the SSK's original order of proof–suspicion–hint, around half (55%) of the participants formed the order proof–hint–suspicion. The participants interpreted the category "hint" as stronger and more conclusive than "suspicion". Furthermore, there were enormous differences between the participants.

Such differing meanings for verbal quantities of risk – in this case, verbal labels for the presentation of conclusiveness – are not only evident among lay people. In a study by Jablonowski (1994), experts (risk managers) were interviewed about the meanings they ascribed to various statements of frequency.[53] Great variations were also found here in the evaluation of terms of probability labels.

Information about uncertainties in risk assessment leads to different appraisals. For some, this leads to an increase of trust in the source of information supplied, while others consider this as a sign of incompetence and dishonesty (cf. Johnson and Slovic 1995). In a part of an experiment in the same study, the presentation of confidence intervals for risk estimates were shown to lead to a higher risk perception than when point estimates were used. This also showed that uncertainties expressed as confidence intervals also led to higher levels of concern.

In another study of the perception and evaluation of uncertainty in risk assessments (Johnson and Slovic 1998), the authors found that the upper limit of defined intervals was seen to be the most credible estimation. Furthermore, the source of uncertainty was seen to lie not in the nature of the issue in question but mostly to be ascribed to social factors (the self-interest of the experts, or incompetence). Practically half the interviewees believed that the authorities only communicated a risk when it was serious. About half the interviewees preferred an "either–or" communication, i.e., a clear message of the existence or nonexistence of a risk.

The findings of Johnson (2003 b) largely replicate these results. These findings showed that general attitudes and evaluations of the information provider had an influence on the assessment of the uncertainty. Those who were more mistrustful were more inclined to consider the statement of uncertainty as a sign of dishonesty and incompetence.

The severity of a damaging event as a factor influencing the perception of a risk was examined in the study of Weber and Hilton (1990). More serious events led to a higher level of probability. Possibly the seriousness of events influenced the assessment of probabilities as a "worry effect", i.e., the more serious events draw the attention to the potential of higher degrees of probability.

This suggests as a consequence that verbal representations of quantities should be complemented by quantitative statements. When summary assessments such as evidence, suspicion, etc., are used, their meaning should be pre-tested. Uncer-

---

52) What was judged was the characterization of evidence by SSK (2001) for the effects on health of electromagnetic fields.

53) Rare – very unlikely – unlikely – somewhat unlikely – likely – frequent – extremely likely.

tainties should be communicated in a simple manner, for example with the help of verbal characterization.

4.3.5
Comparisons: What Influence Does the Nature of the Comparison Have on the Comparison?

The comparison of objects (including risks) to form a ranking can be conducted in a number of ways, e.g., through direct ranking, comparison of pairs, or comparison of threesomes (Von Winterfeldt and Edwards 1986). These procedures differ with regard to the cognitive effort and thus to the necessary prerequisites. They are, however, dependent upon acceptance and other basic requirements.

In the literature of risk comparisons a series of experiments can be found which give information about which problems CRA can deal with in its formation of comparisons. In this respect there are a multitude of empirical and theoretical assessments of individual comparison procedures that cannot be listed here. For this reference should be made to the specialist literature, e.g., Keeney and Raiffa (1976) and Von Winterfeldt and Edwards (1986).

A first problem of comparison results from the reluctance to consider alternatives. The investigations of Baron and Spranca (1997) as well as Ritov and Baron (1999), which examine the influence of protected values (PVs) on the formation of preferences and the consideration of alternatives, are pertinent here.[54] By this term is meant absolute attitudes to values that are so strong as to bar considerations. Such PVs are moral assessments of activities that ignore consequences (one may not do X, however minor the consequences). Thus the ranking of the magnitudes of consequences is ignored (a small potential for damage is as bad as a large potential for damage). Furthermore, such PVs play a larger part in the assessment of initiating action (a change in the status quo) than when refraining from action (retention of the status quo). While the context of detection of a PV is not identical with the circumstances of a risk ranking within a CRA, such PVs can lead to conflicts, for instance, when risks for endangered species and for human life are weighed against each other. This can make comparisons, in particular cost–benefit comparisons, difficult if not impossible (Fiske and Tetlock 1997).

It is also evident that the presentation of comparison options has an influence. If options are assessed separately, they are more readily weighed up than when options are considered simultaneously (Bazerman et al. 1999; Hsee 1996). It seems that participants in the case of simultaneous comparison evaluate as they believe they should; in the case of a sequential procedure they evaluate according to their factual preferences (cf. Irwin and Baron 2001).

It also makes a difference as to how the options are compared. A direct rating is oriented more on values than the method of willingness to pay (Irwin and Baron

---

[54] As a rule the participants in the experiments on PV are brought to a moral dilemma. They have to decide whether a therapy that costs five people their lives but saves the lives of 500 can be used.

2001). The two methods lead to different orders of preferences. When the method of willingness to pay is used, the mode of comparison plays a role. When the sales method is used (for how much money would you dispose of X?) value-related decisions are made. In contrast to this, monetary aspects predominate in the purchase method (how much is X worth?) (Chapman and Johnson 1995; Irwin 1994).

Lastly, it is to be noted that risks are not appraised independently of their benefit in risk assessments (and thus also in risk comparisons). Experiments indicate a connection between perceptions of benefit and of risk (e.g., Fischhoff et al. 1978b; Harding and Eiser 1984; Alhakami 1991; Alhakami and Slovic 1994). Finucane et al. (2000), using the key term "affect heuristic", conducted interesting experiments. They could show that risks are evaluated as lower when higher benefit is seen, and vice versa. Furthermore, the judgments of risk and benefit were more strongly negatively correlated when they had to be made under a high time pressure than those made with unlimited time. In the opinion of the authors the affects related to the source of risk are responsible.

As yet the only empirical investigation into the conduct of a CRA stems from the working group around Granger Morgan of the Carnegie Mellon University (Morgan et al. 2000, 2001; Florig et al. 2001). The participants were risk managers from public authorities and industry and they were asked to conduct rankings. They were provided with lists of nine to eleven risks. First (D1) they had a holistic risk assessment to carry out, after which they were asked to weight the attributes that characterized each risk. From this the multiattribute ranking of risks was calculated (D2). After instruction and an introduction, the participants were gathered together in groups and had to carry out a holistic ranking (D3) and a multiattribute ranking (D4). Then in D5 and D6 individual rankings were again carried out (D5, holistic; D6, multiattribute).

Figure 13 shows the result of the correlation analysis. Between the holistic and the multiattribute rankings one finds a striking correlation (between 0.595 and 0.686). This represents, in the opinion of the authors, a sufficiently convergent validity. The agreement between the participants in the six sessions was also good (correlation between 0.595 and 0.913).

The participants were also happy with the group discussion process before D3 and D4 ($\bar{x}$ = 6.12 on a seven-point scale) as well as with the rankings in D3 and D4 ($\bar{x}$ = 5.8 and 5.41, also on a seven-point scale).

The experiments described here indicate that it is not insignificant as to which people should conduct the comparisons and which method of comparison is used. Thus aversions to rankings can result from stronger attitudes to values. Indeed, our own experiments indicate that, with relatively simple comparisons, different groups assess values and arrive at widely consistent preferences in similar ways. It indeed plays a role whether the comparison is conducted by risks (every risk is assessed alone on all attributes) or by attributes (all risks assessed per attribute). Furthermore, the outcomes are differentiated depending on the chosen procedure. Willingness to pay and direct ratings lead to different orders of preference. Finally, the existing emotional attitudes influence the assessment of risk and benefit, which affects the comparison.

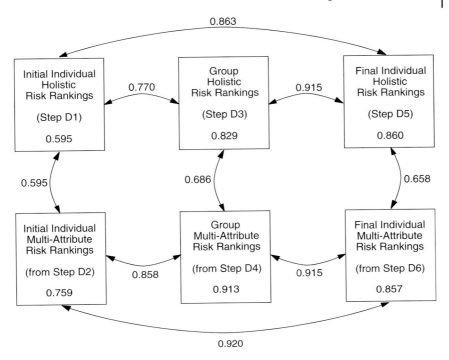

Figure 13
Correlations between risk rankings. (Source Morgan et al. (2001).)

Whichever comparison procedure is chosen, its implementation should be methodologically consistent, fair, and well documented. It is also to be ensured that the procedure selected is understandable and traceable. Thus one arrives at comprehensible and unambiguous characterizations of risk. Wherever possible, several different ranking procedures should be used so that the stability of rankings can be checked.

4.4
CRA Negotiations Under Conflict

The willingness to become involved in a procedure like CRA is not always present. Sometimes opposition or misgivings are expressed. Minard (1994) describes a series of instances.[55]

---

[55] He indicates that a CRA project takes an average of 2–3 years. It takes about 1 year to prepare a CRA and arrive at an appropriate political decision. Then about 6–10 months are needed to perform the procedure. A further year is required for the policy feedback and decision on an action plan. During this time many consultations and tradeoffs are conducted and many decisions taken. (The source at http://www.epa.gov/comp_risk/tool6/part1_2.htm is no longer available.)

A CRA, which involves quite different groups of people, requires agreements. This begins with the establishment of aims, continues with the choice of participants, relates to the determination of the risks to be assessed, their categorization, the choice of attributes, the characterization of risks in regard of these attributes, and finally affects the ranking. Further agreement is required in the development of action plans and their political implementation. Dissent and conflict can occur here.

This includes:

- the insistence of a decision-maker, independently of all other stakeholders, to evaluate all problems separately;
- the anxiety of a decision-maker that, through such a process, they must give up some power;
- the concern that, through the procedure, another position that has not been accepted will be legitimized;
- an absence of knowledge about the design and tools of a consensus-oriented CRA.

In the following we discuss such pitfalls of CRA. They affect all processes of deliberation and assessment in a CRA; this includes the evaluation of one's own position and interests, the evaluation of other parties, and the evaluation of suggestions and findings. Within the discussion we shall follow the literature of negotiation (Bazerman and Lewicki 1983; Bazerman and Neale 1993; Neale and Bazerman 1985; Pruitt and Rubin 1986; Raiffa 1982; Thompson 1990; Thompson and Hastie 1990).

4.4.1
Pitfalls in the Evaluation of One's Own Position and Interests

The most important pitfalls in the evaluation of one's own position and interests are (a) overconfidence, (b) interest-led perceptions, and (c) the illusion of control (see Table 14).

The overestimation of one's own influence (the illusion of control) is the most serious of these pitfalls. One is too certain that the findings will conform to one's own expectations. Overconfidence is particularly noticeable when interim results confirm one's own wishes. This effect is particularly critical, because information that corroborates one's own hypotheses and assessments is perceived selectively and its significance overestimated. Even when evidence of the appropriateness of one's own point of view is lacking, one is inclined to seek only confirmatory information.

Overconfidence can lead to (a) not reaching a possible agreement, because one sticks to one's own position, and (b) the assumption that other participants share one's own position and will go along with assessments which are in one's own interests. In the case of strong conflicts over values, or ties to interests, this attitude can lead to a metaphorical trench warfare in which participants take final stands on their viewpoints, thus preventing integrated solutions. If one is convinced of

Table 14 Biases in the evaluation of one's own party.

Bias	Description	Significance for CRA
Unrealistic optimism	One's own chances in the implementation of interests are too highly estimated	Reduced readiness for concessions and disappointments
The illusion of control	The tendency to ascribe all agreements and arrangements to one's own abilities	Overestimation of one's own negotiating power
Wishful thinking	The probability that a desired event will actually occur is overestimated	Reduced readiness for compromise
Interest-led perspective	Information that confirms one's own point of view is selectively perceived	Overestimation of the appropriateness of one's own position
Seeking confirmation	The tendency to seek information which confirms one's own assumptions and hypotheses but not contradictory information	Ignoring information that would make a reconsideration of one's own position necessary

one's own success, one expects that others must make concessions. If one brings, for instance, representatives of industrial interests and environmental groups together, there is a strong danger that positions, rather than preferences, will be negotiated.

4.4.2
Pitfalls in the Evaluation of Other Parties

One of the most important characteristics of a successful collective evaluation and deliberation process is that the participants are able to take in the perspectives of other parties, i.e., to put themselves in the position of seeing the others' concerns, constraints, interests, and aims. For a CRA in particular, where the representatives of different societal groups who also work together, and so rely on permanently sustainable connections, take part, efficiency and rationality are not the only criteria of good negotiations, but above all the basic principle that working together should improve, or at least not impair, these connections (Fisher and Ury 1981).

The most important misunderstanding and biases working against a flexible and open relationship between the participants are stereotyping, misguided ascriptions of the reasons for behavior, and the overrating of negative impressions (see Table 15).

Through cooperative dealings, each participant forms a picture of the personalities, values, aims, and interests of the other parties. In this way prior experiences are recalled which help to order and structure concrete situations and thus reduce complexity. One of these mechanisms for evaluating other individuals is thinking in stereotypes. Such knowledge of types of people makes it easier to understand even imprecise information about others and from the first impression to deduce

Table 15 Pitfalls in the evaluation of other parties.

Bias	Description	Significance for CRA
Thinking in stereotypes	Hypotheses about the properties, aims, or preferences of a person are derived from an evaluation of the personality type	False assessment of the interests, aims, and purposes
Contrast effect	The greater the perceived difference from one's own attitude, the stronger the tendency to perceive statements or people as contrary to one's own attitude: more contrary than is actually the case	Friend–enemy stereotyping
Consistency effect	Expectations about what will be observed will influence what is actually perceived. Qualities in another, which do not fit one's picture of them, are underestimated or ignored	Inflexible interaction
False consensus effect	When the behavior of another regarding decisions and judgments is similar to one's own, it is perceived as appropriate. When another's behavior deviates, it is regarded as departing from the norm	Black and white perceptions. False assessments and masking of the reasons for others' behavior
Negativity distortion	Negative information more strongly determines the overall judgment of another than positive information	Overrating negative impressions of the interests, aims, and purposes of others
Effect of the first impression	The first impression or first statement is that which is best retained in the memory	Overrating of the first impression

hypotheses about the particularities of their personality, their characteristic qualities, or preferences. However, it can also lead to a false estimation of the other person. The first impression is often overrated. On the other hand, when parties already know each other it is remarkably difficult to make one's assumptions about the intended strategies and interests of other participants more flexible, because one assumes that others will always behave in the same way. The consistency effect additionally helps to maintain stereotyping, because qualities that do not fit the expected picture are suppressed and ignored. The danger of a stereotyped judgment lies in mutual misunderstanding.

The nearer another's attitude is perceived to be to one's own, the greater is the tendency to play down differences. However, the contrast effect leads to polarization. The greater the perceived difference, the more strongly the two positions are blown up into contraries. This tendency to friend–foe stereotyping can have the effect that an existing disparity is overrated and a possible consensus thus not arrived at. Such false estimations can lead to misunderstandings, which promote positional conflicts, leading to the formation of factions and hindering integrated solutions.

False hypotheses concerning the reasons for behavior can have similarly serious effects. To put oneself in another's place also means perceiving situational influences, which provoke particular behavior. Running a "hard" communication strategy is not necessarily a sign of an inflexible character, which one must also deal with under other circumstances. If one assumes a static personality trait, one underestimates the possibilities for influencing other parties or may overlook pressures to act that other representatives may find themselves under. Only understanding of such situational conditions can lead to a fair dialogue, which then also promises efficient solutions.

4.4.3
Pitfalls in the Evaluation of Suggestions

The most important pitfalls in the evaluation of suggestions are (a) the influence of a zero sum assumption, (b) the handling and assessment of losses and profits, and (c) the danger of positional negotiations so as to protect one's own face (see Table 16).

Experiments in negotiation behavior have shown that participants enter negotiations with prior assumptions. The zero sum assumption is of particular significance here. The parties believe that they have differing, mutually competing, aims. They thus define the CRA as a competition that has winners and losers.

Table 16 Pitfalls in the evaluation of the negotiation process.

Bias	Description	Significance for CRA
Zero sum assumption	The assumption that the gains of one party are losses for the other party	Predominance of a winner–loser strategy
Framing effect	When people can choose between a certain and an uncertain alternative, they usually choose the certain alternative, if gains are concerned. They choose the uncertain alternative, if losses are concerned	If the suggested compromise is perceived as a "win" situation, the readiness for acceptance increases and vice versa
Orientation on reduced costs	Previous costs rather than future gains are taken as a basis for decisions	Compromises, which are rational from the point of view of the negotiations, are excluded
Cognitive dissonance	Reviewing previous positions is not in agreement with self-image. Dissonances with self-image are masked or interpreted in a particular way	Room for compromise depends on how strongly one seeks to preserve face

When a suggestion is on the table, one has two alternatives: either to accept this certain alternative or to decide for an uncertain alternative, i.e., to continue negotiating – with the risk that no agreement may be reached, or that the outcome will be worse than that already reached. In this situation the assessment of what has already been reached plays a significant role. It is of consequence whether what is already on the table is regarded as a gain or a loss. If the suggestion is seen as a gain, one is inclined to accept it. If one appraises the suggestion to be a loss, then one is more prepared to take a risk and gamble what is already on the table for an uncertain, but perhaps better, outcome. This evaluative effect operates independently of the content of the suggestion on which one is seeking to agree. The same starting point can lead to differing decisions depending upon whether one regards the offer as positive or negative.

This effect is particularly strong when participating parties discuss positions. When announcing a certain position, a range of anticipated outcomes will be established. So as not to lose face one is obliged to view every suggestion that deviates from this as a loss. An agreement is particularly difficult to achieve in such a case.

The effects can be similarly serious if the participants, while making their decision about whether to accept a suggestion or not, fix their attention on the preceding phases of negotiation. If, in such a consideration, the parties count previous efforts, such as time already invested, trouble, and expense, it may lead to an adherence to one's own conception of objectives. A potential agreement is rejected although a consideration of future costs and benefits indicates that it is rational to make concessions. And the more closely one is bound to a previously adopted position, be it through public statements or an extreme opening position which allows little room for concessions, the more likely an escalation of conflict becomes. This strengthens the tendency to defend one's own position. It is, therefore, particularly important that tendencies to tie oneself to fixed positions are always discussed, alternatives pointed out, and flexibility encouraged.

4.4.4
Pitfalls in the Evaluation of Outcomes

The evaluation of the process to achieve results, and the willingness of the participants to identify with the outcomes, is important, because (a) the agreement, whether achieved or not, has to be communicated outwards (to the public) and inwards (to one's own party) and (b) lessons can only be learned when what has been achieved is appropriately appraised. The pitfalls of overrating one's own role, especially in relation to the success of negotiations and the errors of reconstruction, can be obstructive (see Table 17).

If a CRA fails, it is up to the participating parties to show the reasons for this. If a consensus has been achieved in the risk ranking, the assessment of this outcome can define whether one can be seen as a winner or a loser. Because the tendency is to ascribe success to one's own efforts and abilities, the possibility exists, in the case of failure, to attempt to shift responsibility and that every party tries to claim

Table 17 Pitfalls in the evaluation of the outcomes of negotiations.

Bias	Description	Significance for CRA
Faulty recall	The inability to recall the details of events leads to a "logical" reconstruction that can be inexact	Inconsistencies, gaps, and ambiguities of the process of a negotiation are brought in hindsight into a consistent picture
Self-serving biases	The preservation of self-image through the tendency to ascribe successes to one's own credit and failures to external factors	Responsibility for the failure of a negotiation is shifted away
Egocentricity	The tendency to overrate one's own contribution to a collectively achieved outcome	Exaggerated depiction of a winner's position

success for itself. The inclination to ascribe successes to one's own credit often adds to the overrating of one's own contribution. Ascribing negative results for one's own party, for instance, due to the lack of readiness to compromise on the part of others, therefore occurs not least because in hindsight events which support one's own conception are more readily recalled; similarly, dissonant situations are suppressed or treated as insignificant.

This tendency to suppress contradictions and inconsistencies in hindsight is of little help when trying to learn from a clarification of the processes of a CRA. Precisely for this reason it is necessary to have a clear view of one's own contributions, even when in the course of negotiations and when looking back at their outcome these prove to have been misjudgments. Besides, one is in hindsight often not surprised by the outcome that is arrived at. Plausible explanations are easily found. These, however, hinder an examination of one's own contributions with the purpose of working on these and, from this, configuring further dealings with other parties.

The errors and pitfalls described here need not, but can, arise in the course of a CRA. This is especially likely to be the case if a risk ranking is to be equated with a setting of priorities and it is in the interests of specific participants that certain risks are given a high (or, in the opposite case, low) position.

5
Conceptual Framework for an Integrated Comparative Risk Evaluation

5.1
Methodological Problems of a CRA

The examples of CRAs discussed in Chapter 3, together with the discussion of empirical research into judgment and decision behavior in Chapter 4, show that there are two large and salient problem areas for the conduct of CRA projects: (a) dealing with uncertainties which may exist with the underlying risk assessments, and (b) the choice of evaluative criteria which should be taken into account in a CRA. These are both discussed below.

5.1.1
Problem: Uncertainty and Incertitude

The greatest problem for a comparative risk evaluation is the uncertainties that may occur in the analysis of risks. Thus, for instance, Schuhmacher-Wolz et al. (2002) find, in an examination of risk assessments of carcinogenic air pollutants, that even in qualitatively sound studies there are differences of up to two orders of magnitude for the unit risk in the different assessments. Assessments of exposures can also, depending on the measurement and estimation procedures, vary substantially (cf. Swartjes 2002; WHO 2000).[56] In a summary analysis of this problem, Neus et al. (1998, p. 79) come to the conclusion:

> That in unfavorable constellations individual cases of risk assessments may deviate far from the "true" relationships, and that risk assessments cannot be understood as exact risk prognoses. In so far as the compilation of selected examples allows a generalization, the margin of tolerance of the risk estimates lies in general in the order of magnitude of two decimal powers. [Translated from German]

---

56) Cf. also the series of articles Probabilistische Expositionsabschätzung in Umweltmedizin und Verbraucherschutz (Fehr and Mekel 1999).

Comparative Risk Assessment. Holger Schütz, Peter M. Wiedemann,
Wilfried Hennings, Johannes Mertens, and Martin Clauberg
Copyright © 2006 WILEY-VCH Verlag GmbH & Co. KGaA, Weinheim
ISBN 3-527-31667-1

A further problem in the comparison of risks arising from exposure to pollutants lies in the nature of this exposure: does it actually occur in the sense that one must reckon on its happening, or is it an event which logically cannot be excluded that is to be prevented by a great effort? The first case denotes the overwhelming majority of environmental health risks; the second is typical of risks arising from hazardous incidents in a technology with high and well-known potential dangers, such as nuclear energy. in the second case it remains uncertain, whether hazardous incidents identified as occurring very infrequently ever happen, also – or even more – if the uncertainties of the risk assessment are small.[57]

5.1.1.1 Types of Uncertainty

Different concepts are brought to the discussion of the uncertainty of knowledge about risks. The distinction between aleatory and epistemic risk is frequently made (e.g., Pate-Cornell 1996). Aleatory uncertainty denotes a random process (e.g., the fall of dice or the decay of radioactive substances), while epistemic uncertainty describes limited knowledge.

In many cases, however, it is rather a question of the approach than a property of the quantity under consideration that determines whether an uncertainty is aleatory or epistemic. Thus, the varying sensitivity of individuals to a pollutant is, when considering the entire population, an aleatory uncertainty. When one considers the individual person, though, this uncertainty is more epistemic, and the sensitivity of this person can in some circumstances even be more precisely determined with appropriate tests. However, even if the sensitivity of each individual were exactly known, it would still be reasonable to apply a probability distribution to the description of the collective sensitivity, i.e., to describe it as a random (aleatory) process.

Another distinction often met is between variability and (actual) uncertainty (cf. Bogen 1990; National Research Council 1994). The first signifies the heterogeneity within a population (this relates to the aleatory uncertainty mentioned above). It is thus an objective and, in principle, exactly measurable characteristic of nature. The other signifies uncertainty arising from the incomplete knowledge of a particular issue. This uncertainty is not a quality of nature, but a cognitive characteristic, so to speak, of the scientist or the scientific community (this relates to epistemic uncertainty).

The distinction between variability and uncertainty is of practical as well as theoretical interest. While the variability within a population cannot be reduced through additional research, but at best more precisely described, uncertainty can often be reduced through further research. In this connection, the question arises whether the effort of a further reduction of uncertainty is justified for the risk evaluation at issue – a question that can itself be handled methodologically (value of information analysis; cf. Clemen 1991; Morgan and Henrion 1990).

[57] For the difference between an uncertainty and an incertitude, see Section 2.3.

Various taxonomies have been suggested for the more precise characterization of uncertainty (e.g., Bogen 1990; Morgan and Henrion 1990; National Research Council 1994). A distinction frequently deployed is between parameter uncertainty (e.g., errors of measurement or the validity of indicators) and model uncertainty (e.g., the appropriateness of linear models without threshold values in the determination of dose–effect relationships, or the completeness of models).

5.1.1.2 Uncertainties in the Process of Risk Assessment

The complexity of this problem becomes clear when one considers the causes of uncertainty and variability for the various steps of the process of risk assessment (see Table 18).

Table 18 Some causes of uncertainty and variability in the steps of a risk assessment. (After Bogen (1990, pp. 30–32).)

	Uncertainty	Variability
Hazard identification	– Different study results – Different study qualities – Different study types – Extrapolation of available evidence to target human population	– Heterogeneous metabolic competence – Heterogeneous immunological competence – Heterogeneous absolute susceptibility of target site
Dose–response assessment	– Extrapolation of tested doses to human equivalent doses – Definition of "positive responses" in a given study (as above) – Parameter estimation – Model selection for low-dose risk extrapolation	– Heterogeneous metabolic capacity – Heterogeneous immunological capacity – Heterogeneous quantitative susceptibility of target site – Heterogeneous quantitative availability of target site
Exposure assessment	– Contamination scenario characterization (production, distribution, domestic and industrial storage and use, disposal, environmental transport, transformation/decay, geographic bounds, temporal bounds) – Exposure scenario characterization (exposure route identification, exposure dynamics model) – Target population identification	– Heterogeneity in applicable exposure routes (regarding age, sex, occupation, location of residence, behavior) – Heterogeneity in exposure dynamics (regarding age, sex, occupation, location of residence, behavior, genetic constitution, state of health, random processes)
Risk characterization	Component uncertainties: – hazard identification – dose–response assessment – exposure assessment	Component variabilities: – hazard identification – dose–response assessment – exposure assessment

In the past it was most often attempted to "make up for" these uncertainties – above all for the uncertain assessment of dose–effect relationship and exposure assessment – with "worst case" assumptions. This means that the most unfavorable conditions were used in the form of point estimates in the risk assessment. Such a procedure makes, on the one hand, for a high degree of safety, since health risks cannot be underestimated, but it can lead, on the other hand, to a significant overestimation of risks and thus to a misallocation of resources for the protection of health.

Newer approaches based on probabilistic techniques use for the parameter under consideration, instead of "worst case" point estimates, probability distributions, which are calculated (US EPA 1996; Cullen and Frey 1999) with the aid of statistical methods (Monte Carlo analysis and others; cf. Vose 1996). This procedure uses the available information better and arrives, it is hoped, at more realistic risk assessments. However, a series of assumptions must be made with this approach, for instance about the form of the underlying probability distributions, which must themselves be established and which can strongly influence the results. A comprehensive discussion of both procedures with illustrative examples for the problem of exposure evaluation is available in the works of Fehr and Mekel (1999), Heinemeyer (2000) Mekel and Fehr (2000), Mosbach-Schulz (1999), Schümann (2000), and Wintermeyer (1999).

5.1.1.3 Uncertainties in the Risk Assessment of Technical Plants

When the risk assessment of technical plants (e.g., chemical plants, nuclear power stations, dams) is examined, a further aspect arises: uncertainties in the assessment of the risks of hazardous incidents. Besides the risks associated with the normal operation of plants, the risks of hazardous incidents must also be considered. Concerning the risks associated with the normal operation of plants, the important question is about the environmental burdens created and the possible resultant environmental damage and harm to health. In the case of hazardous incidents, the issue is against which events, due to their plausibility and consequences, measures must be taken, and whether a remaining risk of a still possible hazardous incident is acceptable. Here the characteristic uncertainty is whether a possible – and also plausible – recognized hazardous incident will actually occur within the period that has practical relevance for a risk assessment. In this sense what is no longer practically relevant is, for instance, a period that is significantly longer than the lifetime of the plant in question, or of plants of the type under consideration. Obviously, one does not leave the occurrence of hazardous incidents to chance. Reliable safety installations in technical plants have the effect – according to the way of thinking, in principle justified, of licensing procedures – of "practically excluding" unacceptable hazardous incidents. However, as long as these incidents are not logically excluded, the risk arising from the uncertainty referred to remains.

Technical risk analyses are conducted to determine the frequency of hazardous incidents. They generally inquire into the causes of dangerous deviations from the requirements of safe processes and their consequences. Three methodological approaches may be distinguished here. The first is the intuitive empirical, which relies on, for instance, a brainstorming, the appraisal of operational experiences in accident reports, or working through checklists. The second methodological approach is the inductive procedure. This starts from initiating events and asks what hazardous event sequences can develop from these. Examples are the failure effect analysis and the event tree analysis. The third method, the deductive procedure, begins with undesired plant states and asks how these could have come about. This is equivalent to the principle of a fault tree analysis.

The starting point of a fault tree is the undesirable state of a subsystem ("failure"), for instance of a safety system. The causes that could lead to the situation are deductively determined. The system is broken down into its subfunctions and components; the interrelationship of functions is mapped with the aid of the Boolean operators AND and OR. This is continued to a level on which the component-related frequency of failures is known, from which the frequency of a system failure is calculated. Event trees, in contrast, inquire inductively into the consequences arising from an initiating event depending on the reactions of subsystems (performance of task or failure). Through a combination of event and fault trees, accident scenarios are developed at the end of which stand the conditions of the whole system (the facility or plant) from which the conditional probabilities following an initiating event can be calculated. For each of these plant states, the consequences are determined by modeling the sequences "formation and transport of pollutants in the plant – emission of pollutants – distribution – exposure – effect", under the condition of the given plant state. In this way, the risks associated with complex facilities may be determined.

A principle problem of this procedure is that only risks that are sought for can be found (scope of consideration), and that only risks whose harmful effects are known can be sought.

For the estimation and assessment of uncertainties in the risk analysis of technical facilities it is necessary that the documentation of the risk assessments identifies:
- where the limits of the scope of consideration lie
- at which points subjective judgments are incorporated
- at which points uncertainties exist
- how large these uncertainties are
- where knowledge is lacking
- how large these gaps in knowledge are.

5.1.2
Problem: Evaluative Criteria

The choice of criteria by which a comparative risk evaluation should be performed depends on the nature of the CRA's aim, and especially on the subjects of protection that should be kept in mind in the CRA. The starting points for the derivation of evaluative criteria are the subjects of protection for which the CRA is being performed. As a rule these must be differentiated so that they may be incorporated within the framework of a CRA. This differentiation must lead to criteria for which measuring methods (metrics) are available.

Table 19 Descriptions of Vermont's quality-of-life criteria (US EPA 1993a, p. 2.4–6).

Criteria	Descriptions of criteria
Impacts on aesthetics	Reduced visibility, noise, odors, dust, and other unpleasant sensations, and visual impact from degradation of natural or agricultural landscapes
Economic wellbeing	Higher out-of-pocket expenses fix, replace, or buy items or services (e.g., higher waste disposal fees, cost of replacing a well, higher housing costs), lower income or higher taxes paid because of environmental problems, net jobs lost because of environmental problems, and health-care costs and lost productivity caused by environmental problems
Fairness	Unequal distribution of costs and benefits (e.g., costs and benefits may be economic, health, aesthetic)
Future generations	Shifting the costs (e.g., economic, health risks, environmental damage) of today's activities to people not yet able to vote or not yet born
Peace of mind	Feeling threatened by possible hazards in air or drinking water, or potentially risky structures or facilities (e.g., waste sites, power lines, nuclear plants), and heightened stress caused by urbanization, traffic, etc.
Recreation	Loss of access to recreational lands (public and private), and degraded quality of recreation experience (e.g., spoiled wilderness, fished-out streams)
Sense of community	Rapid growth in population or number of structures, or development that changes the appearance and feel of a town; loss of mutual respect, cooperation, ability, or willingness to solve problems together; individual liberty exercised at the expense of the individual; and loss of Vermont's landscape and the connection between the people and the land

If one considers, for instance, the general subject of protection "human health" one sees that the term health – and therefore also the term health risk – is not as unambiguous that it alone could give rise to evaluative criteria. The definition of health used by the WHO in the preamble to its founding document makes clear that health – and therefore also risk – can be understood as a multidimensional concept: "Health is a circumstance of complete physical, mental and social wellbeing and not merely the absence of illness and infirmity".[58]

The problem for a CRA then arises in how the complex concept of "health" can be operationalized. The aspects of wellbeing mentioned in the WHO definition have usually been incorporated into previous CRAs under the term "quality of life" and dealt with as a separate facet of the assessment alongside "health" defined in the narrower sense (of absence of illness and infirmity) (cf. US EPA 1993a). An example of the spectrum of evaluative criteria for risks to quality of life is given in Table 19.

[58] Preamble to the Constitution of the World Health Organization as adopted by the International Health Conference, New York, 19–22 June 1946; signed on 22 July 1946 by the representatives of 61 states (Official Records of the World Health Organization, no. 2, 100) and entered into force on 7 April 1948.

Regarding the differentiation of evaluative criteria for risks to health (in the narrow sense), the Unfinished Business CRA of the US EPA (1987), for instance, distinguished between carcinogenic and noncarcinogenic effects (see Section 3.2.1). The noncarcinogenic effects were subdivided into three categories of harm:
- *catastrophic effects on health*, e.g., death, reduction of life expectancy, serious physical or mental disabilities, hereditary diseases;
- *serious effects on health*, e.g., dysfunction of organ or nervous systems, as well as behavioral and developmental disorders;
- *adverse effects on health*, e.g., loss of weight, changes to enzymes or reversible organic disorders.

For the category of "adverse effects on health", the question arises when a certain effect is to be regarded as "adverse". According to the WHO definition of "adversity", this is the case when it leads to a reduction of the ability of the organism to function (see Section 2.2). Whether this can be assumed to be for a specific effect is, finally, a question that only experts can judge. They can arrive at entirely differing appraisals of this. For instance, are interruptions to sleep, which are traced by the affected individuals to road noise or a nearby mobile telephone transmitter to be categorized as adverse effects on health or "only" as disturbances of the mental state? Clear criteria need to be set for a CRA here.

If one includes, besides human health, the environment as a subject of protection, then one arrives at an even more complex picture, as the following example for the impact of pesticides shows (Levitan 1997). This lists as criteria:
- Effects on the health of farmers, consumers, and the general public, including particularly sensitive subpopulations.
- Lethal and sublethal effects on flora and fauna.
- Direct and indirect influences on the natural and agricultural ecosystems, including the effects on habitats and sources of nutrition.
- Pollution of soil, air, and water which could have consequences for humans as well as for flora and fauna.
- Societal costs through the remediation of environmental damage.

It is obvious that this list also has to be further differentiated and above all operationalized so as to become useable within the framework of a comparative risk evaluation. What is also interesting about this list is that it subdivides each individual sphere of impact.

A further and adjacent evaluative criterion is the benefit associated with the application of risk-bearing options (technologies, substances, patterns of behavior, etc.). Risks do not arise of their own will, but because one anticipates an individual or societal benefit from the technologies, substances, or patterns of behavior in question. And in fact various studies into the perception of risk have shown that the judgment of risks and benefits are not independent of each other. Rather, there is a negative relationship: high judgments of benefits appear alongside reduced evaluations of risk, and vice versa (cf. Alhakami and Slovic 1994; Finucane et al. 2000; Fischhoff et al. 1978b; Gregory and Mendelsohn 1993; Harding and Eiser

1984; Vlek and Stallen 1981). And naturally, the aspect of benefit plays a central role in the question of acceptance and acceptability.

In the context of a CRA the evaluation of risks should certainly be independent of those of benefits, since these are, for example, also set out in risk–benefit analyses (cf. Merkhofer 1987). It is, after all, about the comparative evaluation of risks. And it is precisely when risk-bearing options are to be evaluated regarding both their risks and their benefits that the assessment must manage to keep both aspects separated, so that utility is not included twice in the evaluation – once implicitly, within the risk evaluation, and a second time explicitly, within the evaluation of benefits.

5.1.3
Consequences for the Conduct of a CRA

The uncertainties affecting the risk assessments that are part of a CRA cannot, of course, be removed from the framework of the CRA. It is, therefore, even more important that these uncertainties are comprehensively described in the risk characterization. Gray (2000) identifies three typical problems or failings of a risk characterization:

- *False precision*. The application of point estimates to the description of a risk – particularly with people who are not conversant with the problems of risk assessment and the resultant uncertainties (regulators, journalists, and the public) – can lead to a deceptive impression of precision and lack of ambiguity in the risk assessment. When conducting a CRA with the participation of stakeholders who are not experts in risk assessment, there is a need within the risk characterization not only for quantitative description of such evaluative uncertainties but also to supplement these with graphs that illustrate and texts that explain.
- *False consistency*. In the context of risk assessments, standard default assumptions are made which are variously plausible in various cases. For instance, in the risk assessment of carcinogenic substances, a linear dose–effect relationship is often taken to arrive at an assessment of carcinogenic potency for low doses. For certain agents (e.g., directly mutagenic chemicals) scientific data in fact suggest such a relationship, but for other agents (e.g., nondirectly mutagenic chemicals) this is questionable. This means that assessments based on such default assumptions may in some cases lead to more precise risk assessments, but in other cases have the consequence of an overestimation of the risk. It is necessary for the risk characterization to explain to what extent similar methods and assumptions in risk assessment lead to similarly plausible risk assessments.
- *Hidden choices*. In the context of risk assessments, assumptions and decisions must often be made to deal with incomplete data and deficits in theoretical understanding. The application of the default assumptions mentioned above, or "worst case" assumptions, are examples of this. Another example is the use of different sampling methods in exposure analysis (cf. IPCS 2000). Depending upon the exposure pathway considered and the point on the exposure pathway where the sample is taken (e.g., quantity of pesticides in cereals in the field ver-

sus quantity of pesticides in the corresponding foodstuff), entirely different values may be obtained. Neither sampling method is inherently superior; it depends in fact on the concrete constraints (time and other resources available) as to which method is used. It is important for risk characterization to make clear which assumptions have been made or which sampling methods chosen so that they can be considered in the assessment of findings.

When one considers the many uncertainties that can occur in a risks assessment – and which can only be discussed in a rudimentary way here – the question of how one should proceed with these uncertainties in the context of a CRA arises.

The next thing to be acknowledged is that a comparative risk evaluation can only ever be as good as the data from the risk assessments on which it is based. And since these uncertainties cannot be removed from the context of a CRA, one should ensure that a comparative risk evaluation uses no risk assessments whose capacities diverge widely. This means that, within a CRA, only risks which are based on conceptually equivalent risk assessments should be compared, i.e., on methods whose application lead to similarly certain or uncertain analyses (Neus et al. 1995, p. 305).

As regards the remaining uncertainties, two problematic facets are to be distinguished: one is the question how uncertainties can be communicated so that participants in a CRA understand what these uncertainties signify. This is above all of importance when stakeholder groups who are not experts in risk assessments are participating.[59] The other is to decide how the uncertainties should be accounted for within the comparative risk evaluation.

The possibilities for the communication of uncertainties depend initially upon whether these may be described quantitatively or merely qualitatively. Quantitative characterizations of the uncertainties associated with an analysis, for instance in the form of confidence intervals for appraised values (average, median), percentages, or probability distributions, exist, and so these can, besides their numerical presentation, be graphically illustrated.[60] Examples of these can be found, for example, in Morgan and Henrion (1990, Ch 9), Thompson and Bloom (2000), and WHO (2000). Even if uncertainties can be only qualitatively described (e.g., "great uncertainty", "rather small uncertainty"), a graphic illustration can also be sought besides the verbal characterization.[61] Care is necessary when using verbal characterizations of uncertainties because, as the investigations into the understanding of verbal characterizations of expressions of probabilities discussed in Section 4.3 show, the meaning ascribed to such descriptions can vary greatly between individuals. In this connection, the findings of Erev and Cohen (1990), also reported in Section 4.3, are of significance; most people prefer to use verbal descriptions when

---

[59] Problems of understanding can also occur among experts of differing professional backgrounds, such as toxicologists and epidemiologists (cf. Bailar and Bailer 1999).

[60] Overviews of the various possibilities for the graphic presentation of quantitative data are available, for instance, in Cleveland (1993) and Tufte (1983).

[61] See also the discussion of the presentation of uncertainties in the WBGU (1998) survey in Appendix 2.

they need to communicate probabilities, but would prefer to be informed of probabilities in a numerical form.

The substantial challenge for CRA lies, however, in the question how uncertainties inherent in the analysis of the risks to be compared may be incorporated within the framework of the comparative risk evaluation. There are here, fundamentally, three possibilities.

1. For the range of values characterized by the uncertainty a specific value is selected. This can be, for example, an average or median, or an upper or lower percentage value. This choice means that the uncertainty can no longer be explicitly considered within the comparison.
2. Various assessment procedures are conducted using different values, for instance firstly with the average, then for a selected percentile. It can then be checked to what extent these produce differing rankings for the various values.[62]
3. Uncertainty is incorporated as a separate assessment aspect (attribute) along with average or median values. In the context of a multiattribute assessment approach, as explained in Section 5.2, this aspect can then be awarded a particular weight and thus the influence of uncertainty on the assessment determined.

Which of these three possibilities is suitable for a CRA can only be decided with regard to the concrete constraints. From a theoretical point of view the second alternative is the best, because it makes transparent the influence of various assumptions about the actual specification of an appraised value on the comparative risk evaluation. It also, however, demands the most effort.

5.2
Methodology of Comparative Risk Evaluation

The methodological framework for the comparison of risks by reference to different evaluative criteria is supplied by the approach of multiattribute decision-making.[63] This approach – also referred to generally as a decision-analytical approach – has in the past been explicitly used in some CRA procedures, above all from the angle of including various stakeholder perspectives (e.g., Apostolakis and Pickett 1998; Bonano et al. 2000; McDaniels et al. 1999; Merkhofer et al. 1997). Implicitly or explicitly, all of the CRA procedures presented in Chapter 3 are based on such a multiattribute structure.

The approach of multiattribute decision-making allows the appraisal of various options in relation to evaluation aspects (attributes) and their integration into a general appraisal, which can then serve as the basis for the decision. This conception can be transferred without difficulty to CRA: the options are either the risks (sources of risk)

---

[62] This corresponds to the performance of a sensitivity analysis, which is examined in Section 5.2.2.7.

[63] There are numerous texts dealing with the theory and practice of multiattribute decision-making. Here we draw largely on Eisenführ and Weber (1993), Jungermann et al. (1998), Keeney (1992), von Winterfeldt and Edwards (1986), and Yoon and Hwang (1995).

to be assessed or specific risk-bearing alternatives. In the first instance, various risks are appraised according to one or more evaluation aspect(s). As a result one obtains a ranking of risks that reflects their importance or seriousness in the view of the assessors. In the second instance, there is a selection of the best of various alternatives, which are appraised with reference to one or more attribute(s).

In this the decision-making is based on two different types of information: one of these is knowledge or expectations about the consequences or effects that may arise from the respective options; the other resorts to decision-making about evaluations. These can, for instance, relate to the assessment of the gravity, desirability, acceptability, etc., of specific consequences or effects. The evaluations can also relate to the significance assigned to various evaluative dimensions.

One could then emphasize that decisions are based on facts and values. In the context of comparative risk evaluations the facts are supplied through the procedure of risk assessment. Naturally – as has been and is repeatedly pointed out (e.g., Shrader-Frechette 1995) – the procedures of risk assessment are themselves based on numerous evaluations. For example, the ends to which a risk assessment is conducted must be selected. Measuring methods and experiment designs must be chosen. Guidelines regarding the quality of data required must be set, and so on. These are all decisions, which finally rest on the evaluations of the participating scientists. However – and this is an important difference from other evaluations – good reasons must as a rule be cited for these decisions and they must also be justified in the critical debate between scientists. These evaluations require justification (even when this justification lies, in many cases, in the background of scientific history and is incomprehensible to nonexperts).

In the following the distinction between quantitative and qualitative approaches is examined. The process of multiattribute evaluations will then be dealt with. This is begun with a single assessor. The special problems of comparative risk evaluations with several stakeholders are then discussed in a separate section.

5.2.1
Multiattribute Risk Evaluation: Quantitative and Qualitative Approaches

Fundamentally, the structure of a multiattribute risk evaluation is very simple: there are a series of attributes $(x_1...x_n)$ for which, in the assessor's view, the estimation u of a given quantity of m risks – or, more generally, options – is of significance. These attributes can be of varying importance to the evaluator and the risks can obtain different evaluations for the attributes (see Table 20).

Table 20 The structure of multiattribute decision-making.

	Attribute 1	Attribute 2	...	Attribute n	Conclusion
Risk 1	$u(x_{11})$	$u(x_{12})$	...	$u(x_{1n})$	?
Risk 2	$u(x_{21})$	$u(x_{22})$	...	$u(x_{1n})$	?
⋮	⋮	⋮	⋮	⋮	⋮
Risk m	$u(x_{m1})$	$u(x_{m2})$	...	$u(x_{mn})$	?

A prerequisite of such a multiattribute evaluation is that all the risks can be evaluated for the respective attributes. This evaluation can occur either quantitatively or qualitatively. Quantitative evaluations are based on quantitative risk assessments and their underlying risk metrics, e.g., "number of deaths" or "relative risk". On this basis cardinally scaled risk values may be calculated which permit an exact quantitative comparison (e.g., risk 1 is three times as high as risk 2).

As shown in Section 5.1, quantitative risk assessments are frequently beset with large uncertainties, so it can make sense to gather assessments with similar values and uncertainties into categories. For instance, the frequency of damaging events can be ordered into the categories of "rare", "occasional", and "often". Or a category that relates to the gravity of a risk can be designated "low", "medium", or "high". Because such a categorization of quantitative data is always accompanied by a loss of information, it appears that this is only appropriate when the uncertainties within the risk assessments are so large that they may no longer be meaningfully characterized by statistical measures such as mean or variance. In extreme cases there is a complete absence of quantitative data so that only a rough categorical analysis of risks is possible.

A qualitative comparative risk evaluation can also appear to be indicated on practical grounds, e.g., so as to simplify the procedure for lay people. In every such case, however, the question arises whether on the basis of such reduced data a comparative risk evaluation actually has any information value.

A special variant of qualitative risk assessments are those in which the findings are presented in the form of categories although the risk assessments upon which they are based are of an entirely quantitative nature. Above all in the context of risk regulation risk classes are often formed for which specific measures are then scheduled. Examples of this include the classification of carcinogenic chemicals in workplaces in the classificatory schema of the MAK Commission (Greim and Reuter 2001), or the classification of carcinogenic air pollutants in the emission limitation classes of TA Air (Schuhmacher-Wolz et al. 2002; see Section 3.2.6). The basis for such a classification into risk classes here is an evaluation of the evidence for the carcinogenicity of the substance, which deals with quantitative risk assessments as well as an appraisal of the quality of available data. In the WBGU survey *Welt im Wandel – Strategien zur Bewältigung globaler Umweltrisiken* (WBGU 1998) the risk assessments were gathered into categories which no longer indicated an ordinal relation to each other but referred directly to particular risk management strategies.[64]

[64] See Section 3.1.3 and Appendix 2.

5.2.2
Conduct of a Multiattribute Evaluation

Seven steps may be distinguished in a multiattribute evaluation process:[65]
1. problem structuring
2. selection and categorization of risks to be considered
3. attribute generation and structuring
4. appraisal of risks for attributes
5. attribute weighting
6. aggregation of individual evaluations and (provisional) ranking
7. sensitivity analysis.

There are now a range of computer programs that support the performance of multiattribute evaluation processes and decision-making. These programs are very helpful – at least when a quantitative multiattribute assessment is to be conducted – above all in the appraisal of options (risks) for attributes, in the weighting of attributes, in the aggregation of individual evaluations, and in the sensitivity analysis, because they contain processes which, for instance, elicit weightings or value functions (utility functions) and can perform consistency tests. And, in practice, a sensitivity analysis without computer support is extremely laborious to perform. However, many programs now support the selection of options and the generation and structuring of attributes. A current overview is to be found in Maxwell (2002). Keefer et al. (2003) give an overview of development perspectives of decision support methods, and criteria for the selection and evaluation of decision support software are discussed in Rohrmann and Schütz (1993).

5.2.2.1 Problem Structuring
At the beginning of a CRA, a series of specifications and stipulations must be made so as to bring the general aim of the comparative risk evaluation on to an operational level:
- Which risks should be considered?
- For which geographical area (national, regional, local) and for which population should risks be compared?
- Which timeframe should be considered: present risks or future risks?
- Who are the evaluators and which evaluations should they make (weighting of attributes, appraisals of the attributes)?
- How should the evaluation of risks proceed, qualitatively or quantitatively? This depends above all on the following aspects:
  - Is a quantitative evaluation, given the state of the data, i.e., because of uncertainties that may be present in the risk assessments, possible and meaningful?
  - Who are the addressees of the comparative risk evaluation, i.e., for whom are the findings of the CRA intended, and what purpose do they have?
  - How far should the aggregation of evaluations for attributes go?

[65] Following the SMART method from Edwards (1977) and Edwards and Barron (1994).

In many cases this first step is largely determined by the context in which a CRA is to be prepared and the concrete requirements resulting from this.

5.2.2.2 Selection and Categorization of Risks to be Considered

This step is also given in many cases by the context in which a CRA is performed: when, for instance, in the framework of a comparative assessment a list of chemical agents has already been compiled.

If a list of risks to be considered is not yet available, it can be gained, for instance, from surveys. It should be noted generally that the comparison of a large number of risks quickly runs up against practical limitations (cf. Morgan et al. 2000). It therefore makes sense to categorize risks according to, for example, particular problem fields such as activity fields (nutrition, etc.), which are then subdivided into problem fields (e.g., foodstuffs, internal air) in a further structuring step. At the lowest level of these problem fields, specific lead indicators must then be assigned (e.g., pesticide burden, acrylamide burden, or burdens of radon or passive smoking) which indicate the sources of risks to be compared. These must be selected so that the significant risk problems in the problem field are covered.

Even at this early stage of the structuring, numerous decisions are to be made which can influence the outcome of the CRA strongly. So it must be decided, for instance, whether for the problem field of internal air other indicators (e.g., formaldehyde) should be considered besides radon and passive smoking. It may easily be appreciated that the choice of categories and indicators can have a decisive influence on the findings of the CRA.

It is also to be noted when structuring the risks that only risks for which conceptually equivalent risk assessments are available, i.e., those based on methods whose application leads to similarly certain or uncertain assessments, are selected (Neus et al. 1995, p. 305).[66]

5.2.2.3 Attribute Generation and Structuring

Attributes constitute those aspects of the risk evaluation that are of significance in the eyes of the evaluator. These can be, for instance, the subjects of protection affected (e.g., human health, the environment, quality of life) or the nature of damage or harm (death, chronic impairment, etc.). For instance, the attributes shown in Figure 14 were used in the New Jersey Comparative Risk project (NJDEP 2003). Here, three spheres for comparative risk evaluation were distinguished, human health, ecological criteria, and socioeconomic criteria, which were each characterized by three attributes. In the project, these three spheres were not gathered into an overall evaluation, but remain as the outcomes of three separate comparative risk evaluations (which also lead to different series of rankings).

Whether single spheres such as human health, ecological, or socioeconomic aspects are considered separately within the framework of a CRA or – see the example shown in Figure 15 – gathered into an overall evaluation depends on the aim of the CRA. When aggregating different spheres into an overall evaluation, though,

---

[66] See Section 5.1.3.

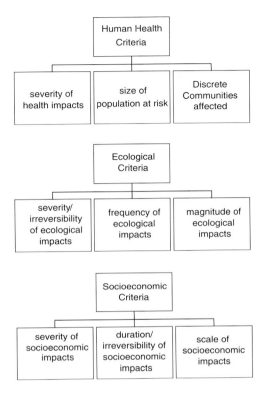

Figure 14
Example of attributes in the framework of a CRA. (After NJDEP (2003))

these should in every case also be considered separately so as to understand which possible rankings of risks they yield. Within the framework of a multiattribute approach this is easily possible.

The selected attributes must – and this is equally the case for both quantitative and qualitative assessment methods – fulfill various requirements to be of use in the framework of a multiattribute evaluation procedure. Criteria for these are:

- *Completeness.* It must be ensured that all the aspects significant to the evaluation are considered. This is of particular significance with multistakeholder evaluations (see below).
- *Nonredundancy.* The attributes should not overlap in their subject, because such overlaps can lead to a situation in which an attribute obtains a greater influence over the decision-making than it deserves.
- *Measurability.* The attributes must be operationalized in such a way that it is possible to measure the occurrence of risks for the attributes. Firstly, such measurements should be as apt as possible, which means that they should measure those aspects that are most important to the evaluator. Secondly, the measurements should be as unambiguous as possible, which means that vagueness in the measurements should be as low as possible.

- *Independence.* The appraisal of risks from one attribute should be possible independently of the appraisal from the other attributes. A change in the appraisal from one attribute ought not, therefore, lead to a change in one or more other attribute(s). This independence of preferences should not be confused with an empirically given independence in a concrete case (statistical independence). The attributes may well in fact correlate for given alternatives, i.e., the character of one attribute relates to that of another. The independence of preferences simply requires that the evaluations must in principle be possible independently of each other.

It often makes sense to organize these attributes in the form of hierarchies of aims or value trees. Forming such a hierarchy helps the assessor to be clear about the various assessment aspects and their relations to each other (see Figure 15). What is important is that a metric is available for the terminal attributes, i.e., the attributes on the lowest level[67] with which the corresponding risks may be appraised.

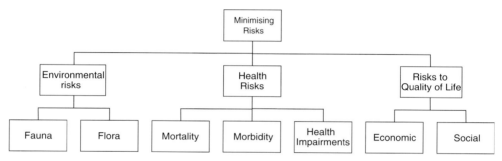

Figure 15
Example of a hierarchical attribute structure.

The choice of terminal attributes therefore also depends on the risks to be assessed, or, more precisely, on the level of resolution at which the risks are considered. The risk metrics to be found in Appendix 1 give an overview of the bandwidth of available measuring methods. In cases in which no data using such risk metrics are available, indicators of risk can be used. However, for these a plausible case must be made as to why these are relevant to the evaluation of the risk.

Naturally, the problem of the measurability of terminal attributes occurs mostly in quantitative risk evaluations. However, this is also true in a reduced form in qualitative evaluations, because it must also be possible to carry out an evaluation of the risks through classifying them in categories for the relevant attribute.

Attribute structures can be developed either top-down or bottom-up.[68] In the former, one seeks firstly to specify general risk fields, aims, or values that are relevant

---

67) The use of the term attribute in the literature is not entirely unambiguous. Sometimes the terminal attributes are referred to as attributes and those above as goals. We always refer to attributes regardless of the level.

68) The development of attribute structures is discussed in detail in Keeney (1992).

to the risks to be considered. In the next step these general concepts are differentiated until a hierarchical structure, at the bottom end of which are the attributes that can be appraised, is given. The bottom-up method begins with a collection of the most concrete possible appraisal aspects for the risks. These are then gathered under generic terms and thus hierarchically structured.

Regarding the number of attributes to be considered, the common sense rule applies: as many attributes as are necessary (to grasp all the aspects relevant to the assessment), but as few as possible (to not lose the overview). It is also to be kept in mind that a large number of attributes is mostly more difficult to manage in qualitative than in quantitative evaluations. It is also to be noted that it makes little sense to include evaluation criteria for which no useable data exist. Thus, one may wish, for instance, to take the aspect of "effect on future generations" as a significant evaluation criterion in the framework of a CRA. However, when no scientific assessments of this are available, or when the analyses contain great uncertainties, the inclusion of this criterion is not likely to lead to progress.

5.2.2.4 Appraisal of Risks for the Attributes

The appraisal of risks for the attributes is the task of experts, because this does not involve questions of evaluation but the determination of facts, i.e., an appraisal on the basis of risk assessments – even when these are affected by more or less substantial uncertainties.

There are significant differences between qualitative and quantitative methods of evaluation in the type of appraisal. In qualitative methods this appraisal proceeds by classification into categories. The number of appraisal categories used for the characterization of attributes depends largely on practical requirements. There are no general guidelines for this. It certainly makes sense, however, not to plan too many categories. Three categories are often used, as in the example from the New Jersey Comparative Risk project (NJDEP 2003). Here the categories "low", "middle", and "high" were used to characterize the attributes of "severity of specified health effects at current levels of exposure", "size of population at significant risk for each health effect", and "overall risk ranking (as a function of severity and population affected integrating across health effect)". As well as this a dichotomous category with the characteristics "yes" or "no" for the measurement of the attribute "are there discrete communities at elevated risk?" was used. Table 21 shows some examples of appraisals made by experts for the respective fields on the basis of risk characterizations.

One type of semiquantitative risk appraisal presents rating scales in which the extent of the appraisal of the attribute is expressed numerically. The categorical appraisals presented in Table 21 were additionally expressed in the form of ratings in the New Jersey Comparative Risk project (see Table 22).[69] While with such nu-

---

[69] The instruction for the rating here reads (NJDEP 2003, Appendix 4): *Severity of specified health effects at current levels of exposure:* 1–5 with 1 being least severe. *Size of population at significant risk for each health effect:* 1–5 with 1 being smallest. *Are there discrete communities at elevated risk:* 1–5 with 1 being the lowest probability that there are discrete communities at elevated risk. *Overall risk ranking (as a function of severity and population effected integrating across health effect):* 1–5 with 1 being the lowest overall risk.

Table 21 An example of a qualitative multiattribute risk evaluation.[70]

Source of risk	Severity of specified health effects at current levels of exposure	Size of population at significant risk for each health effect	Are there discrete communities at elevated risk?	Overall risk ranking
Airborne pathogens	Low	Unknown	Yes	Low/middle
Arsenic	Middle	Middle	Yes	Middle
Cadmium	Low	Middle	Yes	Low
Chromium – total				High/middle
Chromium – cancer	High	High/middle	Yes	(n.a.)
Chromium – allergic dermatitis	Middle	Low/middle	Yes	Middle
Chromium – noncarcinogenic ingestion	Low	High/middle	Yes	Low/middle
Dioxins/furans – cancer	High	High/middle	Yes	High/middle
Endocrine disruptors	Low/middle	Middle	Yes	Middle
Greenhouse gases	Low	High	Yes	Low
Ground-level ozone	Middle	High	Yes	High
Pesticides – food	Middle	Middle	No	Middle
Radon	High	High	Yes	High
Second-hand tobacco smoke	High	High	No	High
UV radiation – total				Middle
UV radiation – skin	High	Middle	Yes	Middle
UV radiation – eyes	High	High	No	Middle
UV radiation – immune system	Low	Low	No	Low

merical values the single evaluations of the attributes may be calculatively aggregated to overall values, such calculations are only meaningful if it can be assumed that these appraisals at least have interval scale levels (that is, equidistant scale points). Without this (substantial) assumption, rating scales have a merely ordinal level and therefore express no more than the verbal categories that express ordinal relations.

What is to be remarked here is that verbal characterizations of attribute characteristics in the form of categories may seem at first sight to be simpler than rating scales or the quantitative appraisals presented in the following. This doesn't mean, however, that they necessarily represent the appraisal more reliably. Thus, a comparison of the categorical, verbal estimations in Table 21 and the numerical rankings in Table 22 shows that the classifications of verbal characterizations and numerical values are not always consistent, either for a source of risk (appraised by an expert) or between sources of risk (appraised by various experts). For instance, for ground-level ozone the category "high" is classified as either value 4 or value 5, and for arsenic and cadmium the category "middle" is assigned once as value 3 and

[70] Chosen from NJDEP (2003, Appendix 4).

Table 22 Numerical evaluations.

Source of risk	Severity of specified health effects at current levels of exposure	Size of population at significant risk for each health effect	Are there discrete communities at elevated risk?	Overall risk ranking
Airborne pathogens	1	Unknown	n.a. (yes)	2
Arsenic	3	3	3 (yes)	3
Cadmium	1	2	3 (yes)	2
Chromium – total				4
Chromium – cancer	5	4	4 (yes)	4
Chromium – allergic dermatitis	3	2	3 (yes)	3
Chromium – noncarcinogenic ingestion	2	3	3 (yes)	2
Dioxins/furans – cancer	5	4	3 (yes)	4
Endocrine disruptors	2.5	2.5	5 (yes)	2.5
Greenhouse gases	1	4	3 (yes)	2
Noise – effects on circulation	2	2	5	2
Ground-level ozone	4	5	5 (yes)	4
Pesticides – food	2	3	n.a. (no)	3
Radon	5	5	5 (yes)	5
Second-hand tobacco smoke	5	4	n.a. (no)	5
UV radiation – total				3
UV radiation – skin	3	3	5 (yes)	3
UV radiation – eyes	3	4	1 (no)	3
UV radiation – immune system	2	2	1 (no)	2

once as value 2. If one recalls the discrepancies reported in Section 4.3 between the estimation of probabilities in verbal and numerical forms, this inconsistency will not come as a surprise.

The quantitative appraisal of risks for the respective attributes is based on the assessments that exist in the risk metric with which the attribute is measured. To make the attribute appraisals in the various metrics comparable, they are rescaled using a weighting function (referred to formally as a value or utility function)[71] onto a common evaluation scale. This evaluation scale comprises the interval [0;1], in which 0 is the lowest and 1 the highest value concerning the designated attribute.

Monotonically increasing and monotonically decreasing weighting functions are to be distinguished from nonmonotonic weighting functions. A monotonically increasing weighting function is chosen when a higher occurrence of the attribute also signifies a higher value, e.g., if the attribute reads "extent of reduction of pollu-

[71] We do not distinguish here between utility and value functions. This distinction is usually made in the literature of decision analysis, so that utility is used under uncertainty and value under certainty. In the construction of utility functions the attitudes towards risk are considered, while this is not the case in the construction of value functions. For a criticism of this distinction, see von Winterfeldt and Edwards (1986, Ch 7.1).

tant emissions". Monotonically decreasing weighting functions are used for attributes with which damages or costs are assessed, e.g., "number of deaths" for "mortality". Nonmonotonic weighting functions are used when the best occurrence of the attribute does not lie at the end of the attribute scale.

The construction of weighting functions can be very laborious.[72] In many cases, however, one can assume that linear weighting functions present an adequately good approximation of the actual weighting function.[73]

A linear rescaling of the attribute assessments on the assessment scale can, for instance, be conducted as follows. The value 0 is assigned to the worst value of risks for the attribute and the value 1 to the best value of risks from the attribute. Then the value $u_i(x_{ij})$ of the risk i can be calculated for the attribute j thus:

$$u_i(x_{ij}) = \frac{x_{ij} - x_{ij}^-}{x_{ij}^+ - x_{ij}^-} \tag{5.1}$$

where risk x_{ij}^- is the worst value of risk for the attribute and x_{ij}^+ the best value of risk for the attribute.

If in doubt, it must be checked whether the weighting function of the decision-makers is really linear. For attributes that, for instance, use a monetary scale, a utility function is often nonlinear – above all if the scale spans a large measurement range. The gain from the utility of money is greater in the lower range than in the upper range of the scale. This appears as a concave utility function (see Figure 16).

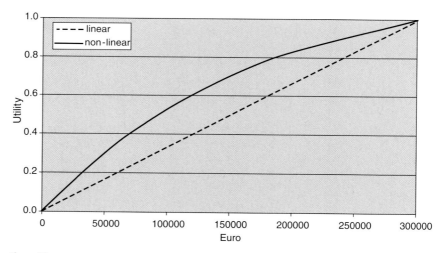

Figure 16
Example of a linear and a nonlinear utility function of money.

72) Various methods for the construction of utility functions are described in detail in Eisenführ and Weber (1993), Jungermann et al. (1998), and von Winterfeldt and Edwards (1986).

73) Studies show that changes in the weighting influence the decision more strongly than the form (linear versus nonlinear) of the utility function (cf. Hobbs 1986; von Winterfeldt and Edward 1986).

If, in the framework of a CRA, only risks are considered, the attribute scales can as a rule be constructed so that larger risks are represented through higher occurrences on the scale. Thus monotonically increasing weighting functions can be used for the assessment of risks and the evaluation scale becomes a scale of risk.

5.2.2.5 Attribute Weighting

Attribute weighting makes explicit the differing importance that various attributes can have for the evaluator and incorporates them in the comparative evaluation. The procedure is different for qualitative and quantitative risk evaluations.

Because in qualitative risk evaluations the appraisal of risks for attributes is available merely on the level of an ordinal scale, it makes little sense to conduct differentiated numerical weightings. If different levels of importance should be accorded the attributes, it is sufficient to weight the attributes in an ordinal manner, so as to bring them into a ranking. The requirements for the application of noncompensatory evaluation procedures are thus given.

With quantitative comparisons the attribute weighting proceeds by mapping the importance on the interval [0;1]. In this way direct and indirect weighting methods can be distinguished.

Weighting methods using ranks belong to the direct method. In this, the n attributes of the assessor are firstly ordered according to their importance. The most important attribute is given the rank order number $r_1 = 1$, the second most important $r_2 = 2$, and so forth. The rank sum weight can then be calculated using the following formula:

$$w_j = \frac{(n - r_j + 1)}{\sum_{k=1}^{n}(n - r_k + 1)} \tag{5.2}$$

The rank sum weighting maps the ranks linearly (i.e., with equal distance) on the interval [0;1], where the sum of weights is equal to 1.

Another method of calculation is the reciprocal rank weighting:

$$w_j = \frac{1/r_j}{\sum_{k=1}^{n}(1/r_k)} \tag{5.3}$$

Here the ranks are likewise mapped on the interval [0;1] and the sum of weights is equal to 1. However, the weights of the ranks do not decrease linearly, but in relation of the placing of their rank to the first rank, i.e., the weight of the second rank is only half that of the first; that of the third rank only a third of the first, and so forth. With this method of calculation the high ranks gain a special weight.

Rank sum weights are also relatively easy to perform for a great number of attributes, but the instructions for calculating the transformation of the information on the ordinal scale in the interval scale level of the weights allows the evaluator no possibility to express the differences in importance of individual attributes beyond the bare formation of a rank.

A simple method which does allow this is point weighting. Here a specified number of points (e.g., 100) are distributed among the attributes by the assessor so that together they reflect the relative importance of the attributes. The standardization on the interval [0;1] proceeds through the division of points by the total number of points (e.g., 100) distributed. Point weighting is simple so long as there are only a few attributes to be weighted. If there are a larger number of attributes to weight, this method quickly becomes unclear and a consistent weighting is difficult to perform.

A central point of attribute weighting is that the estimation of the importance of attributes must be conducted for the particular decision context (above all in view of the other attributes). It does not, therefore, deal with the abstract or general importance of an attribute. Above all, the bandwidth of possible values of the attribute must also be allowed for.

One method, which incorporates this aspect into the imposition of weights, is swing weighting.[74] This weighting method is an example of an indirect method of weighting attributes. Firstly, the best and worst conceivable value is set for each of the n attributes. A hypothetical risk is then supposed which has the worst occurrence on all attributes: $R_0 = (x_1^-, ..., x_j^-, ..., x_n^-)$. The evaluator then selects the attribute that in his or her view the most important, and its value is changed from worst to best ("swing"). A new hypothetical risk is designed which contains this altered attribute (x_j^+): $R_1 = (..., x_j^+, ...)$. The previous step is then repeated with the attribute that is in the evaluator's view the next changed from worst to best value. From this, one obtains a second hypothetical risk, R_2. This method proceeds until all the attributes have been altered to their best value. One then obtains the "least worst" hypothetical risk, R_n. These n risks form a ranking. For each risk R_j the evaluator then marks on a scale of 0 to 100 how large the relative improvement t_j is due to the swing of the attribute from x_j^- to x_j^+. The weights for the n attributes can subsequently be calculated where values for all t_j are standardized on the interval [0;1], i.e., $w_j = t_j / \Sigma t_n$.

If a large number of attributes need to be incorporated into the evaluation, the imposition of weights can be simplified with a hierarchical attribute structure. Then in the first instance the attributes within each branch and on each level of the attribute structure can be weighted against each other. For the attribute structure presented in Figure 15 this means that the attributes on the lower level, flora and fauna, the attributes of mortality, morbidity, and adverse effects as well as social damage and economic damage, must be weighted against each other. On the upper level the three attributes environment, health, and quality of life are then weighted against each other. Admittedly the weights so derived are in the first instance relative weights. The absolute weights required for the multiattribute assessment are obtained by multiplying the weights of the terminal attributes with the weights of the attributes that are on the upper level of each respective branch.

---

[74] See Jungermann et al. (1998, p. 355).

For instance, if the attributes of mortality, morbidity and adverse effects have received the relative weights of 0.5, 0.3, and 0.2, and the attribute health the weight of 0.4, this gives an absolute weight for mortality of $0.5 \times 0.4 = 0.20$, for morbidity of $0.3 \times 0.4 = 0.12$, and for adverse effects of $0.2 \times 0.4 = 0.08$.

The absolute weights of the terminal attributes can thus deviate very widely from the relative weights imposed. Generally, the weights of the terminal attributes can be as high as the weight of the upper attribute of the branch to which they belong. The absolute weights as a rule turn out significantly lower than the relative weights imposed. For example, Figure 17 shows that the adverse effects on health show by far the lowest weight of all attributes, and mortality has only a little more weight than the social aspects of quality of life. After calculating absolute weights, therefore, the extent to which these actually mirror the assessor's estimation of the importance of attributes should always be checked.

Figure 17 also makes clear that the difference between the weights of attributes can be slight. Because the individual weights must add up to a sum of 1 (or 100%), the "elbow room" for individual attributes quickly becomes narrow when there is a larger number of terminal attributes. A purely calculative approach allows arbitrarily differentiated attribute weights to be specified, but their practical explanatory power is indeed questionable. The practical use of highly differentiated attribute weights in multiattribute risk evaluations is also often limited because no exact scientific analyses for the appraisal of risks for attributes are available, so only analyses with an uncertainty (i.e., a certain bandwidth) can be given. Depending on which value is taken from the field of analysis for the risk appraisal, marginally different attributes in the weighting lead to changes in the conclusions and possibly also in the ranking of risks (see also Section 5.2.2.7). These remarks are not in-

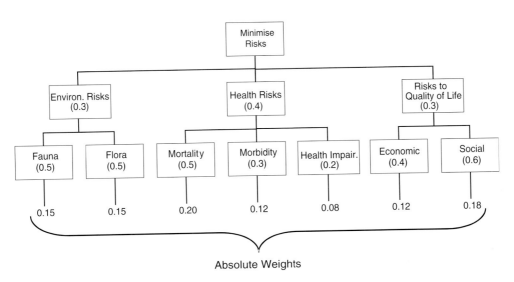

Figure 17
Example of attribute weighting (relative weights in parentheses).

tended to call into question the significance of attribute weighting, and certainly not the importance of transparency in weightings. One must, however, be aware that one cannot offset inexactitudes in the risk assessment through precision in the attribute weighting. In cases where there are large uncertainties in the risk assessments, it is most often more sensible to limit the CRA to a few important attributes.

5.2.2.6 Aggregation of Individual Evaluations and (Provisional) Ranking

There are two types of method by which the individual evaluations of risks for attributes can be gathered into an overall evaluation for all risk: compensatory and noncompensatory methods. Compensatory methods allow evaluations of risks for attributes to be offset against each other. For example, a high evaluation of a risk for attribute A can be compensated by a low evaluation of risk for attribute B. This possibility does not exist with noncompensatory methods. If a risk does not attain a fixed threshold for one or more attribute(s), it is not chosen, regardless of how it performs for other attributes.

The qualitative evaluation of risks with the aid of categories does not permit compensatory evaluation methods, since the categories stand in merely ordinal relation to each other and cannot be offset against each other. In this case, then, noncompensatory methods must be adhered to.

Noncompensatory methods

There are numerous noncompensatory methods of evaluation.[75] In the present connection the following are of interest:
- *Dominance.* The risk which is estimated to be at least as high as all other risks with all attributes and which is estimated to be higher with at least one attribute is chosen.
- *Lexicographic order.* The risk that has the highest estimate with the most important attribute is chosen. If several risks should have equally high estimates, the second most important attribute is considered in the same way, etc.
- *Conjunction.* The risk that attains a determined threshold with all attributes is selected.
- *Disjunction.* The risk that attains a determined threshold with at least one attribute is selected.

Decisions are relatively easily made following noncompensatory rules. Fundamentally it involves ordering the risks according to the appropriate method. It is obvious that the simplicity of the noncompensatory method is its greatest weakness in the context of risk evaluations. It is unclear what should happen if several risks or risk categories perform equally well under the application of a rule. Either a new evaluation using further accepted attributes must be conducted or an arbitrary choice must be made. This last is hardly acceptable in the context of risk evaluations. It is more important to recognize that the demand for a balanced risk

---

[75] For details, see Jungermann et al. (1998, Ch 4.4).

evaluation can hardly be met through the (over-)simplification of an evaluation task characterized by many evaluation aspects either (1) to a yes/no decision in regards to meeting a threshold value (in conjunction and disjunction) or (2) to a one-dimensional sequential comparisons (for lexicographic ordering).

Compensatory methods
Although there are also various approaches within compensatory methods, in practical application one theory takes center stage: multiattribute utility theory (MAUT). Following the decision-making rules of MAUT, the decision-maker selects the option that has the greatest total utility. Total utility is the sum of the weighted individual utilities of an option for the attributes.[76] Within the context of a comparative risk evaluation, this means that the overall evaluation of a risk is the sum of the weighted individual evaluations of the risks in regards to the attributes. This can be expressed in the following formula:

$$u(R_i) = \sum_{j=1}^{n} w_j u(x_{ij}) \qquad (5.4)$$

where $u(R_i)$ represents the overall evaluation of risk R_i over the n attributes, w_j represents the weight of the attributes x_j (with $w_1 + w_2 + ... + w_n = 1$), and $u(x_{ij})$ the evaluation of risk R_i for the attribute x_j.

Table 23 illustrates the evaluation matrix underlying this additive multiattribute evaluation. The result is that one obtains a conclusion for every risk, which can be compared with those of the other risks and thus leads to a ranking.

Table 23 General structure of a multiattribute evaluation.

	Attribute 1 Weight 1	Attribute 2 Weight 2	...	Attribute n Weight n	Σ
Risk 1	$w_1 u(x_{11})$	$w_2 u(x_{12})$	...	$w_n u(x_{1n})$	$\sum_{j=1}^{n} w_j u(x_{1j})$
Risk 2	$w_1 u(x_{21})$	$w_2 u(x_{12})$	...	$w_n u(x_{2n})$	$\sum_{j=1}^{n} w_j u(x_{2j})$
⋮	⋮	⋮	⋮	⋮	⋮
Risk m	$w_1 u(x_{m1})$	$w_2 u(x_{m2})$	...	$w_n u(x_{mn})$	$\sum_{j=1}^{n} w_j u(x_{mj})$

5.2.2.7 Sensitivity Analysis

The purpose of a sensitivity analysis is to determine how robust the (provisional) ranking is against changes to the evaluations of the attributes and/or changes in the attribute weightings.

---
[76] Where no weighting is given, all attributes are equally weighted.

It is rarely possible for the appraisal of risks for attributes to be so unambiguous that no deviations are conceivable. Uncertainties often attend the characterization of risks for attributes. Mostly the appraisal actually used represents just one – perhaps the most probable or plausible – of the appraisals possible. In the context of sensitivity analyses, the appraisals of risks for attributes are systematically altered and a new conclusion made. As a rule this is done unidimensionally, i.e., the appraisal of a single attribute is altered, so that the relationship between the alteration of the evaluation and the conclusion remains traceable. Naturally, however, multidimensional sensitivity analyses, i.e., the simultaneous alteration of several attribute appraisals, are in principle possible. Thus, sensitivity analyses also deliver information about how large the uncertainties associated with the assessment of individual effects underpinning attribute appraisals may be without causing changes in the ranking.

Sensitivity analyses can be conducted not only of attribute appraisals but also of attribute weightings. Weighing up the importance of individual attributes is often difficult for the evaluator, particularly where attributes relating to elementary values are in question. One example is the consideration of the importance of consequences for the health of children, adults, and old people. Sensitivity analyses can here show if, or to what extent, differences in weightings lead to changes in the ranking. This application of sensitivity analyses takes on a particular prominence in the context of multistakeholder evaluations, because it can test whether different estimations of the importance of single attributes have an impact on the ranking (see Section 5.2.2.8).

It is to be noted regarding sensitivity analyses of weightings that not only the weighting of an attribute can be altered to observe what changes may ensue. Due to the standardization of weights to the value 1, it must also be established how the other attribute weights must be altered when one attribute weight is varied. One possibility is to alter the weights of the other attributes so that they continue to stand in the same relation to each other.

The performance of sensitivity analyses can be very time-consuming. This depends, naturally, on the number of attributes to be tested and the fineness of the variations. Nonetheless – or rather, therefore – this step is of great significance within a multiattribute risk evaluation, because it shows to what extent the assumptions made in the appraisal of risks for attributes and also the valuations which define the weightings of the attributes together influence the risk evaluation.

5.2.2.8 Multiattribute Risk Evaluation with Several Stakeholders

Fundamentally, the structure of evaluation problems with several evaluators is no different from those with single evaluators. In other words, these likewise assess risks for various attributes and aggregate these attribute evaluations into an overall evaluation. The decisive difference regarding the necessary extension of the multiattribute concept lies in the amalgamation of (possibly) differing perspectives on the problem, differing risks, differing attribute structures and attribute weightings, and different evaluations of risks from the attributes. Figure 18 illustrates this problem.

Multiattribute multistakeholder evaluations are only meaningful as cooperative procedures, i.e., with all participants in agreement about performing the steps (tasks) of a comparative risk evaluation discussed above to arrive at a common piece of work in a group process. This also means that criteria and rules must be established about how to deal with differences in evaluation or procedural dissent.

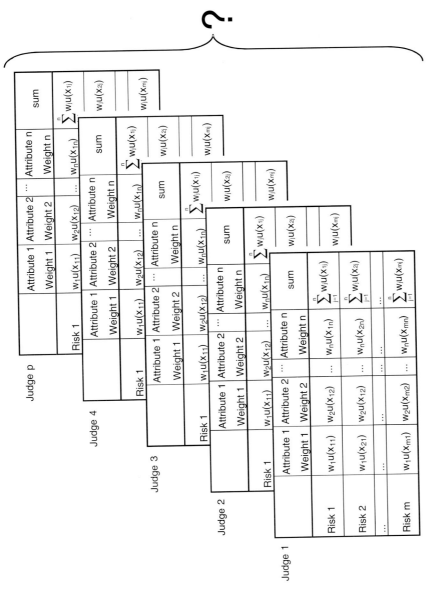

Figure 18
Multiattribute multistakeholder evaluation.

Fundamentally the following criteria must be observed (US EPA 1993a, pp. 2.1–21):

- *Consistency.* The procedure should permit a consistent comparison of risks. It should ensure, for instance, that the criteria selected can be applied to all risks and are observed by all participants. This cannot be guaranteed in methods that, for example, allow secret voting.
- *Fairness.* The procedure should be open and transparent, so that potential distortions in the process of the comparative risk evaluation can be discovered by the participants and discussed.
- *Documentation.* Every step in the assessment process should be documented so that decisions and agreements may also be traced later and appraised by outsiders. This contributes substantially to transparent risk comparisons and makes the communication of findings easier.

In the concrete performance of this group process several variations are possible. At one extreme the entire evaluation process may be conducted together, e.g., as a workshop. At the other extreme the individual steps are performed separately by the stakeholders and their findings presented and discussed at a meeting. While attribute weightings and risk evaluations on the attributes can be quite different for the stakeholders, a prerequisite for a multistakeholder assessment is that agreement is reached regarding the problem and attribute structures.

Firstly, then, a common view of the problem must be developed, i.e., agreement about the structure of the problem must be reached. This involves, on the one hand, the risks and the space-time horizon to be considered. It must also be made clear who deals with which evaluations. On the other hand, a common attribute structure must be developed.

The starting point for such a common attribute structure can be the attribute structures of the individual stakeholders; or the attribute structure is developed by the participants in a group process (bottom-up or top-down). It is important for both procedures that an attribute structure is arrived at which incorporates all the evaluative aspects that the various participants consider significant. Attributes that individual stakeholders consider insignificant can be left included. These can simply be weighted with 0 by these stakeholders and thus be effectively excluded from their evaluation.

The determination of the attribute weights by the stakeholders can proceed in three ways. (a) In the best case, all the stakeholders, perhaps after thorough discussions, can agree on a common weighting. (b) The various weightings are mathematically aggregated, e.g., through averaging, to a common weighting. This procedure is only likely to meet with the agreement of all the participants if the weightings of the stakeholders are close together. Otherwise this determination of the weighting difference will merely mask real differences in evaluations and lead to unacceptable decisions or rankings. (c) The different weightings are retained – at least initially – side by side. This leads in fact to separate evaluations by the various stakeholders.

The appraisal of risks for the attributes by experts (where appropriate selected by the stakeholders) proceeds in the same way as the attribute weighting: through agreement on a common evaluation or through mathematical aggregation or through a provisional retention of the different appraisals.

Naturally, the same ranking of risks can result despite differences in weightings or in risk appraisals. In such a case the differences would not be so serious as to have an impact on the ranking. Should such differences lead to divergent risk rankings, however, sensitivity analyses can be used to test which changes to the attribute weightings or risk appraisals can lead to a unified ranking. The purpose of sensitivity analyses in this circumstance is not to check how robust the ranking is against alterations in the attribute weightings or risk appraisals, but how slight such alterations need to be to allow the attainment of a unified ranking for all stakeholders.

Whether the changes in attribute weightings or risk appraisals necessary to reach a unified ranking are acceptable to the stakeholders cannot, of course, be determined from the sensitivity analysis. The stakeholders must here decide whether these alterations are compatible with their initial evaluations.

6
The Practical Implementation of CRA

Four factors need to be taken into consideration for the practical implementation of CRA. Firstly, the limits of comparison must be defined. This also involves dealing with meaningful possibilities for comparative evaluations of risk. Secondly, it must be made clear who should compare risks; there are actors to be defined and their roles to be determined. Thirdly, the sequential order of CRA procedures must be specified. Fourthly, the communication problems between participants must be minimized.

6.1
Limits of Comparability

One of the objections repeatedly raised against risk comparisons involves the question whether, and if so which, risks can be compared. One of the caveats of the German Risk Commission regarding the comparison of risks is that only comparables should be compared (German Risk Commission 2003, p. 33).[77] What, though, is comparable?

The comparison of the risks of alternative decisions is beyond dispute. This has a long tradition in the field of technology. For instance, locations or technical options for the power industry (Keeney 1980; see also Kunreuther and Linnerooth 1983), or the waste industry (Powell 1996), and also variations for the transport of hazardous goods (rail versus road) may be compared. Likewise, it is reasonable in medicine to compare various therapies regarding their risks (and their benefit). However, whether quite different risks, e.g., radon, nuclear power, global climate change, and internal air pollution, can be compared is, in contrast, doubtful (see Femers 1993). This problem is discussed below.

There are, thus, two issues: the first is to do with different types of knowledge about risks; the second to do with the contrast between factual and hypothetical risks.

---

77) "In comparative risk assessment it must in particular be checked that only comparables are compared" (German Risk Commission 2003, Appendix 4, 34 [Translated from German]).

Comparative Risk Assessment. Holger Schütz, Peter M. Wiedemann,
Wilfried Hennings, Johannes Mertens, and Martin Clauberg
Copyright © 2006 WILEY-VCH Verlag GmbH & Co. KGaA, Weinheim
ISBN 3-527-31667-1

The knowledge problem

Knowledge about an actual or putative risk may be of completely diverse quality. We have elsewhere (Wiedemann et al. 2002 b) proposed a classification of risk knowledge and distinguished types according to state of knowledge: besides known risks there are unknown risks and unclear risks.

A known risk exists if the potential danger (the hazard), the dose–effect curve, and the exposure of the population have been sufficiently well investigated and thus appropriate knowledge is available. Here a complete risk characterization can be conducted. Unclear risks are characterized by a substantial lack of information. It is not clear whether a potential danger exists. Information about the dose–effect relationship is also missing; likewise no details are available about critical exposures. A risk characterization may only be qualitatively conducted. Unknown risks, finally, are those for which no specific details are available.

Principally two possibilities lend themselves to dealing with analysis uncertainties. Either the uncertainty is – as WBGU (1998) did – treated as an attribute in the comparative risk evaluation,[78] or risk classes are differentiated and separate risk comparisons performed for these.

Our opinion is, however, that it is not always productive to assimilate the assessment uncertainty as an attribute among others in risk comparisons in an attempt to incorporate the knowledge problem. It is unclear how this "uncertainty attribute" should be scaled and weighted. It is indeed possible to specify gradations (little, medium, much knowledge; see also Florig et al. 2001) and to weight these classes of knowledge. A substantial problem, however, remains: how can the unknown (does X represent a hazard?) be accounted for alongside other attributes (e.g., number of individuals affected)? The comparison of, for instance, radon and electromagnetic fields is hardly possible in this way. The potential danger of radon is known; the dose–effect relationship is indeed not beyond dispute, but one proceeds from a linear relationship without threshold. The critical knowledge deficit here lies only with the exposure. In contrast, it is unclear whether, in exposures to electromagnetic fields below the limiting value, a potential danger even exists (cf. SSK 2001). A risk comparison fails here because in the case of radon the number of cases of cancer to be anticipated each year can be calculated, while this remains entirely unclear in the case of electromagnetic fields. That the unarguable ubiquity of electromagnetic fields can have no significance for a risk comparison is obvious. The fact that electromagnetic fields are to be found everywhere is of no help in estimating the risk as long it is unclear whether electromagnetic fields even present a potential danger.

Risk comparisons between different classes of knowledge throw up significant difficulties. It is our opinion that risks of divergent knowledge classes (known versus unclear hazard potential) should not be compared. As opposed to the case of known risks, the evaluation of unclear risks firstly involves the performance of a systematic and transparent comparative evaluation of evidence, which means the ascertainment of the state of scientific evidence and if necessary the clarification of what the differ-

---

[78] See example in Appendix 2.

ent scientific estimations are based on. The major challenge for an evaluation of unclear risks lies in its practical organization and conduct (see Appendix 3).

Fact and hypotheticality

Technical risks are often expressed as damage–frequency diagrams (also known as Farmer diagrams), which show what damage is to be expected with what frequency. Risk comparisons between various technological options (e.g., nuclear power versus fossil fuels) then face the problem of how to compare risks which, on one hand, arise from a combination of very low frequencies and very high damages and, on the other, relatively high frequencies with lower damages. To put this in another way, the issue is how to weight potential catastrophes.[79] It is quite another matter when risks are viewed from the perspective of biological endpoints.

The distinction between these two, the technical and the biological, concepts of risk becomes relevant when risks arising from the conventional operations of facilities with low emissions are compared with those arising from hazardous incidents.

Technology-related and biological risks can only be compared with difficulty, because they proceed from quite different understandings of exposure (cf. Uth 1991). Technical risks refer the probabilities of hazardous incidents with the possibility of exposure to a pollutant. The relation between exposure to the pollutant and the resultant harm can then be deterministic, stochastic, or unknown. In contrast, biological risks refer in the first place to an existing exposure. They characterize the probability of adverse effects due to an exposure to the pollutant. In contrast to the risks of hazardous incidents, which deal with merely potential exposures, real exposures are here available.

There are thus risks, which by logical standards exist, but which in all probability will never become reality, or have only an extremely low probability of becoming reality. Typical examples are very severe hazardous incidents in technical facilities which have been identified using probabilistic risk analyses and whose probability of occurrence, so calculated, is very small. In Germany a procedure of this kind has only been practiced in the case of nuclear power; the term "residual risk" has been coined for such a risk. Large chemical plants and dams also belong, due to their potential hazards, to the technical facilities in which hazardous incidents must be prevented with great dependability, and for which accordingly low hazardous incident probabilities exist. Extremely low probabilities also appear for risks arising from extremely rare natural catastrophes, such as earthquakes of a strength that in Germany are conceivable from a geological viewpoint but which are historically unattestable.

In technical plants for which such risks have been ascertained, measures (such as construction designs or safety systems) should guarantee that specific events do not occur. Then "all considerations of damage, all calculations and all factual-technical measures for the prevention of damage are to be sought not in the sphere of fact but in the sphere of hypothesis" (Häfele 1990, p. 33).

[79] In this it would be consistently presupposed that in the case of fossil fuels a hazardous incident analysis is to be conducted using a hypothetical worst case scenario.

The distinction between factual and hypothetical risks can be made clear with the following thought experiment. It is all the same to an individual whether during their lifetime they develop cancer due to a hazardous incident with a probability of 1:1 million or whether they develop cancer due to a continuous exposure with the same probability. The issue appears entirely different, though, when viewed on the level of collective risk. In the case of the given exposure, an individual risk of 1:1 million in Germany leads to an expectation of 80 real deaths. In the case of the hazardous incident risk, no such statement of the really occurring deaths can be made.[80]

We therefore suggest that hazardous incident risks are strictly distinguished from risks lying in the sphere of the factual. In the first case, the probability of occurrence of a damage-causing event is in question, in the second the effect of an existing pollutant. From this it follows that the mixing of potential and real circumstances of exposure should be avoided in risk comparisons.

6.2
Goals of a Comparative Evaluation of Risks

There are a series of possibilities for the comparison of risks. Among these the classic CRA is just a special case of comparative analysis. If one proceeds from the phases of risk regulation (cf. German Risk Commission 2003), from preliminaries through to risk management, one finds various reasonable variants. Table 24 gives an overview of these.

6.2.1
Preliminary Analysis

In the preliminary phase three risk-related comparisons are meaningful. One is the scanning of new technical fields to discover new potential risks. The second deals with comparative studies of risk perception. The third involves the prioritization of substances or other sources of risk to which a risk assessment should be applied.

The scanning of new technical fields for the early recognition of potential risks is based predominantly on the application of qualitative methods, in particular the scenario technique. Such scenarios must be in line with the framework of risk assessment. Central questions are:
- What new technologies, materials, products, etc., are coming towards us? How will these be developed on a quantity basis?
- In which fields of activity do they have an impact?
- Which potential stressors/risk fields can be connected with them?
- Which population groups will be exposed to these stressors?

[80] This in no way minimizes the significance of the risk of hazardous incidents, but shows that they must be dealt with separately.

Table 24 Possibilities for comparative analyses.

	Aim	Focus of comparison	Model	Involvement
Preliminary analysis	Evaluation of new potential risks	Comparison of new fields of technology	Scenario technique and MAUT	Information, consultation
	Analysis of the potential for public mobilization of new technology	Comparison of risk perceptions	Psychometric approach	Participation
	Evaluation of the urgency of a risk assessment	Comparison of substances of possible concern	MAUT	Information, consultation
Risk assessment	Assessment of the evidence of a risk	Placing evidence on an evidence assessment scale. Comparison of various assessments of evidence	Evidence-based medicine. Consensus model	Information
Risk evaluation	Evaluation of the magnitude of damage potential	Comparison of different substances regarding their potency, e.g., their endocrine-modulating effect	Ranking	Information
	Evaluation of exposure to a pollutant	Comparison of sources of exposure	Ranking	Information, consultation
	Evaluation of the vulnerability of a population	Children versus adults	Comparison of risk assessments	Information, consultation
	Evaluation of risks	Programmatic risk comparisons up to largely homogeneous risk lists	MAUT	Participation
Risk management	Comparison of technologies	Cost-benefit comparison of the choice of management options, risk tradeoff analyses, comparison of potential for public mobilization, locations	MAUT, cost-benefit analysis	Information, consultation

The core problem here is the recognition of signals. Creativity techniques as well as forecasting methods are available for these difficult steps. Creativity techniques aid thinking outside the usual channels. For instance, morphological methods are systematic analytical procedures which structure the finding of ideas through ordered thinking, whereas the explorative method of brainstorming works through the open expression of ideas where picking up and further developing the ideas of other participants is encouraged. Delphi procedures offer another approach. The core of this procedure is to gather the opinions of experts, e.g., about the future development of technology, in several rounds. Using a formalized questionnaire, a typically large number of experts will evaluate specific theses. In one or more feedback rounds the experts subsequently gain the opportunity to reflect on and, as the case may be, modify their judgments in the light of their colleagues' estimations.

For the comparative evaluation of future risks one uses methods that enable the most transparent and comprehensible weighting of individual evaluation criteria, such as the newness or the spatial dimensions of a future field of risk. The approach of multiattribute evaluation described in Section 5.2 can be applied here.

To obtain a "thick" description of possible fields of risk evaluated as significant, the scenario method can be used. Here the goal is the development of several alternative futures. A scenario is a set of characteristics of influencing factors, which describe a future situation as well as the events leading up to it and their sequence in time. Using various assumptions regarding the direction and strength in which these factors can work and a consistent combination of such assumptions for various influencing factors, various scenarios can be developed. By the analysis of influences, fields of influence are identified, factors affecting the fields of influence ascertained and evaluated, and the network between the fields of influence acquired. The driving factors are elaborated in the findings. A participation of stakeholders in terms of information and consultation is sensible here.

For the analysis of the potential of new technologies for public mobilization comparative studies of risk perception can be helpful. There have been such investigations since the 1970s. Such comparisons are multidimensional. In these, people judge numerous sources of risk (e.g., living near a chemical plant, cycling, and nuclear power) in regard to their riskiness. In addition, these sources of risk are judged according to various qualitative dimensions, which are assumed to be of significance in the judgment of risk. Such dimensions include, for instance, the seriousness or gravity of a risk, the voluntariness of the risk, the controllability and the familiarity of the risk, or catastrophic potential associated with the source of risk. The data gained for every source of risk and judgment dimension are aggregated across the respondents in the statistical analysis so that a mean value results for every source of risk and judgment dimension. These data are then evaluated with the help of multivariate analysis procedures such as factorial or regression analyses. Such studies have shown that risk perception depends upon, among others, the following factors (cf. National Research Council 1996, p. 147; Wiedemann et al. 2000):

- The risk is immediately comprehensible and can be connected to existing beliefs.

- It possesses dramatic qualities: the pollutant is characterized by a high "dread" factor (creating very strong fear and revulsion) and by the involuntary exposure of a large number of people.
- The individuals are identifiable and culprits can quickly be found.
- The societal distribution of risks and benefits is seen to be unfair.
- Similarly structured risks have in the past led to controversy.
- The institutions responsible are not trusted or lack societal legitimization.

By definition, the perspectives of the affected individuals (stakeholders and the public) are at the center of studies of risk perception.

At times, due to the multitude of substances in need of regulation, particularly urgent cases must be identified, i.e., priority substances for which a risk assessment is to be performed (see German Risk Commission 2003). Such *priority assessments* are likewise based on comparisons. Several characteristics are as a rule evaluated here: emission data, exposure characteristics, evidence of toxicity. An example is the DYNAMEC approach of the OSPAR Commission (2000, 2002). Here entries to the list of *Substances of Possible Concern* were made, chiefly on the basis of PBT criteria.[81] Whichever criteria are drawn into the analysis, methodologically the multiattribute assessment approach described in Section 5.2 also provides the foundation for this type of comparative risk evaluation. In general, participation of the public occurs here by providing information and consultation.

6.2.2
Risk Assessment

The phase of risk assessment deals with the evaluation of evidence that suggests that an adverse effect is caused by an agent. The problem here can be that various experts or groups of experts arrive at differing risk assessments. While this is nothing unusual in science, it can often create misgivings among the public. Since in practice such dissent between experts can frequently be the cause of conflict about risks,[82] the comparison of differing evaluations of evidence is an essential variant of CRA.

Fundamentally, differing experts' appraisals can arise at every stage of the risk assessment process: when ascertaining a potential hazard, when analyzing the dose–effect relationship and the exposure, as well as during the characterization of risks. The question of whether a potential hazard actually exists, however, is the one that finds a particular resonance with the public. It therefore seems advisable to clarify the differences and similarities in experts' appraisals with the aid of a transparent procedure in a systematic comparison of such risk evaluations. The first proposals for the conduct and practical implementation of such a comparative procedure have been made by the California EMF project (Neutra et al. 2002) and by Wiedemann et al. (2002a).

---

81) PBT = persistence, bioaccumulation, toxicity.
82) One can in fact state that the differing risk appraisals of experts are a major cause of all societal conflicts about risks (cf. Mazur 1981; Peters 1993).

It is essential to gather divergent experts around a table and work through the arguments for and against the existence of a potential hazard in a comprehensible fashion (see the example in Appendix 3). This basic discussion structure can serve as the starting point for a clarification of the differences in evidence assessment between the various experts and expert groups. Such a procedure comprises three main steps, which the participating experts and expert groups must take:
1. the appraisal of the findings regarding the quality of existing scientific studies;
2. on this basis the formulation of arguments for and against the existence of a causal connection;
3. the final (qualitative or quantitative) evaluation of evidence.

There are a number of guidelines for the selection and assessment of scientific evidence.[83] So, for instance, *Cochrane Reviewers Handbook* provides detailed information on all facets of a systematic search and assessment of the literature in the medical field, especially for randomized controlled studies.[84] The handbook of the British NHS Centre for Reviews and Dissemination is similarly detailed.

The possibility for the participation of stakeholders and the public in this evaluation of evidence lies in the evaluation of the persuasiveness of the arguments for and against.

Another variation of evidence comparison is an evidence evaluation scale such as those available from the IARC. Another example is the classification schema of the German Commission for Radiation Protection (SSK), which distinguishes, in its appraisal of the evidence of the effects on health of high- and low-frequency electromagnetic fields, between scientific evidence, scientifically founded suspicion, and scientific indication (SSK 2001, p. 7). These are defined as follows:

- *Scientific evidence* of a connection between an adverse effect on health and electromagnetic fields exists if scientific studies by independent research groups can reproducibly demonstrate this connection and the general scientific view supports the existence of a causal relationship.
- A *scientifically founded suspicion* of a connection between an adverse effect on health and electromagnetic fields exists when the confirmed findings of scientific experiments demonstrate a connection but the generality of scientific experiments does not entirely support the existence of a causal connection. The extent of scientific suspicion is based on the number and consistence of scientific works available.

[83] For a general overview, see Cooper, H. and Hedges, L.V. (eds.) (1994) *The Handbook of Research Synthesis*. New York: Russell Sage Foundation, Part III.
[84] Clarke and Oxman (2002).

- *Scientific indication exists when several experiments indicate a connection between an adverse effect on health and electromagnetic fields but are not supported either by independently confirming experiments or by the general scientific view.*   [Translated from German]

Such categories are helpful for a comparative evaluation of evidence because they offer an ordinal scale by which the evidence for adverse effects of various substances may be compared. The evidence classification regarding specified endpoints (notably carcinogenicity) for a large number of pollutants have for some time been available in the form of publicly accessible (via the Internet) databanks so that such rankings can be accessed or created without great data recall efforts. Notable in this respect are the databanks of the IARC,[85] the IRIS system,[86] of the US EPA, and the scorecard system[87] of the Environmental Defense Organization. In the case of evidence scaling such as this, the participation of the public is best undertaken by information and consultation.

6.2.3
Risk Evaluation

Four possibilities for comparisons exist in the phase of risk evaluation: (a) the evaluation of the potency of a pollutant, (b) the evaluation of exposure to a pollutant, (c) the evaluation of the vulnerability of populations, and (d) the comparative evaluation of various risks.

The comparative evaluation of the potency of a pollutant may involve the assessment of its endocrine-modulating effect (cf. Gaido et al. 1997). The comparison of sources of exposure is important, for instance, in the selection of minimizing measures, such as when they deal with the exclusion or limiting of those sources that make the large contributions to exposures. Finally the comparison of the vulnerability of parts of the population in the determination of precautionary measures is significant. Here the interesting question is whether such groups exist and how much more sensitively they react to the pollutant than the normal population. These variations of comparative risk evaluation require special expertise and provide only limited opportunity for the participation of the public.

The central variant of comparative risk assessment, which proceeds with the participation of lay people, is the programmatic comparative risk assessment presented in Sections 3.1.1 and 3.2.1. While there is as yet no practical experience in Germany, as opposed to lessons learned garnered in other countries, it is quite conceivable to conduct such a CRA in the context of the German Federal Action Programme Environment and Health or similar interagency initiatives. Further advice is to be found in Appendix 2.

[85] http://www.iarc.fr/
[86] http://www.epa.gov/iris/index.html
[87] http://www.scorecard.org/

6.2.4
Risk Management

Comparative risk evaluations here relate to the action alternatives for managing a risk. Typically these deal either with the question of which option is best with regard to a desired benefit or with which option for risk reduction is most economical from the point of view of cost efficiency. Examples of the first are the choice of suitable locations for a waste disposal site or the comparative evaluation of power supply systems (see Sections 3.2.4 and 3.2.5).

Examples of the second type are the assessments of measures for the reduction of health risks in relation to their estimated costs per life saved (e.g., Ramsberg and Sjöberg 1997; Tengs et al. 1995).[88] As mentioned in Section 2.7, it must be noted that in both cases the avoidance or minimization of a risk can itself result in risks among its consequences. Possible displacements of risks should therefore be considered in the evaluation and selection of risk management options in the framework of a risk tradeoff analysis (Graham and Wiener 1997; Presidential Commission 1997, p. 35).

This form of comparative risk evaluation must, besides risk assessments, also incorporate benefit assessments. In fact, other criteria beyond quantifiable risk and benefit assessments are under some circumstances consulted for the decision regarding risk management options. Neus et al. (1995, p. 261) note in this connection:

- the consideration of trends over time (particularly with regard to possible worsening of problem situations);
- the inclusion of central ideas such as the realization of sustainable development;[89]
- the review of the social fairness of action alternatives; [89]
- the availability of legal and institutional room for maneuvering.

This shows that the application of comparative risk evaluations in the context of risk management raises far-reaching methodological (e.g., the incorporation of cost–benefit analyses) problems and even problems of environmental health policy. These go significantly beyond the scope of this project and this book.

---

[88] The catalog of options investigated stretches from measures for raising transport safety through measures for the control of pollutant emissions to medical therapies and precautionary measures.

[89] The GaBE project discussed in Section 3.2.5 provides an example of the incorporation of such aspects.

6.3
Participation Models

A systematic stakeholder analysis precedes the planning of participation. This should determine the following. (a) Who is affected by the CRA? (b) Which spheres/sectors (e.g., business, environmental groups, consumer protection), and also which geographical regions will be represented by whom? (c) Are there predominant risk conflicts? (d) Are there citizens' initiatives in specific regions or problem areas? (e) What interests do the various stakeholders represent? What positions do they have regarding various risk problems, and how do they stand regarding a CRA? (f) How well are the stakeholders organized? What support do they need? (g) What costs will arise?

The participation of the public can be implemented according to various models.[90] Two basic variations may be distinguished: the participation of organized interest groups or of citizens who do not belong to any organized group.[91] A procedure that combines both variations is also conceivable. In Figure 19 five levels of participation are differentiated.

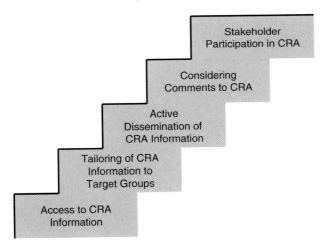

Figure 19
Levels of participation.

The stepped levels arise from the readiness to supply information and the readiness to address the concerns of the public as well as the extent to which citizens are involved in CRA.

90) See Chemical Manufacturers Association (1988), Edison Electric Institute (1994), Canada Standards Association (1996).

91) An example of this is the planning cell approach of Dienel (2002); see also http://www.planungszelle.de/

The *first level* of participation involves access to all the risk information of a CRA. Not only the findings of the CRA are made public, but also all the data and appraisals which underlie the risk assessment and risk evaluation processes.

The *second level* of participation is reached when all the information of the CRA is suitably prepared for public use so that the criteria of comprehensibility, clarity, and conciseness are met. At issue are, in particular, transparent and comprehensible information (e.g., summaries of risk information in question-and-answer format), the availability of evaluation aids, and background information, as well as directions to supporting information.

The *third level* of participation involves an active dissemination of information about the CRA. The CRA project managers do not wait for the requests from the public but deliver information on their own initiative. Central points are: (a) early information about risk assessments and evaluations, (b) definition of target groups for risk information, (c) selection of appropriate communication channels and media for the target groups, (d) publicizing the communication channels to the stakeholders, and (e) testing whether and how the communication is received by the target groups.

The *fourth level* of participation is reached when the CRA project managers, e.g., through focus groups or surveys, collect the concerns of the stakeholders and incorporate them in the planning and performance of the CRA.

Consultations with affected parties are characteristic of the *fifth level*. The CRA project management actively includes public participation in the comparative risk evaluation.

6.4
CRA Participants

A potential CRA participant is anyone who is a concerned citizen, manager, operations staff member, supporting expert, or active user of a CRA project. These various actors operate in different roles in order to contribute to the success of the CRA. The challenge, as in any participatory process, is to balance the need to consider all participants' viewpoints with the practical considerations of organizing a group of individuals who have a role in making or directly influencing CRAs.

CRA project management
The CRA project management is responsible for the planning and performance of the CRA. Its duties include the specification of the CRA's aims, the development of an initial framework for the CRA, as well as the establishment of a steering group.

The project management should move the procedure forward and bring it to a successful conclusion. It coordinates all the planning activities for the successful accomplishment of the CRA. It has the task of maintaining the balance for information, knowledge, or other inequalities. In this respect it must take into account that, besides different conditions of knowledge, the interests of the participants also differ and that different experiential backgrounds and expectations can influence the CRA. Furthermore it is incumbent upon the CRA project management to safeguard the quality of experts' inputs for the risk assessments. It is also responsible for project-related public relations.

Steering group
The steering group offers a practicable possibility of bundling the expert knowledge of various institutions and incorporating this into the planning and performance of the CRA. Such steering groups have proved themselves in other projects.[92]

The steering group consists of members of the organization commissioned to conduct the CRA (e.g., the upper federal agencies) and representatives of other governmental and nongovernmental institutions. It is advisable even at this stage to see that there is some representation of relevant societal groups, or at least that they are informed. Further, experts who possess special knowledge of risk communication and organizing participatory procedures may be consulted. The group serves to support the project management. As a rule it is particularly active in the preparatory phase and advises on the development of the conceptual framework of the CRA.

Citizens' advisory committee
It is the task of a citizens' advisory committee to create, on the basis of the expert evaluations of risks, the risk ranking and to give recommendations about this. In cases in which the participation of the public is particularly critical, it should be considered how additional inputs can be given for the CRA through public hearings, for example through the selection of problem lists for the ranking.

It must in every case be considered anew who should participate on a citizens' advisory committee. There is no formula for this. It is a matter for negotiation, which depends on the circumstances of the location and the aims of the procedure. Besides societal interest groups (environmental groups, patients' groups, consumer protection, industry) there are also affected citizens to consider. To be noted in the choice of participants are the following:
- Have all the important stakeholder groups been incorporated?
- What requirements do the parties demand of the procedure?
- Is a readiness to engage in discussions discernable?
- Are there reasons for any of the parties to obstruct the CRA?
- What incentives can improve their motivation to work with the CRA?

Expert teams
The expert teams gather the data for the risk assessment, evaluate the risks, and can also produce a first ranking, which is later discussed and modified by the citizens' advisory committee. They have to explain in hearings the assumptions underlying their evaluations and the remaining uncertainties. The experts come from state bodies, from private institutions, or from universities and other research establishments.

[92] See Wiedemann et al. (2002); also Renn et al. (2003).

Depending upon which risks are to be assessed from what perspective, determines the different knowledge and skills required.

Addressed stakeholders
The stakeholders addressed by a CRA are all the interested parties. It is therefore advisable to create a list of all groups to be informed about the procedure. Potential participants, interested experts, nongovernmental organizations, policymakers, and all groups directly affected by the findings of the CRA belong to this.

To ensure the CRA's successful beginning, these groups should be informed in good time about the aims and procedures of the CRA. It is advisable to produce a leaflet or short brochure and to develop a presentation, which can be distributed during the various preparatory presentation meetings.

The clarification of the cooperation of the citizens' advisory committee with the expert teams is of particular significance. This begins with the selection of experts. The CRA steering group may justifiably be inclined to assign representatives of the authorities to the expert groups to save costs. However, this can meet with opposition from the citizens' advisory committee, which may prefer the broadest possible expertise and may also wish to have its own choice of experts incorporated. A consensus approach is to be advised in such a situation. It is advisable to concede a right to propose members of the expert teams to the citizens' advisory committee.

A second point requiring clarification involves the distribution of rights and the competence to make decisions between the experts, the planning team, and the citizens' advisory committee. A division between assessment and evaluation tasks has proved worthwhile. Once the list of risk problems to be evaluated has been delineated, the experts take on the necessary work of risk assessment and risk characterization, which form the foundation of the ranking undertaken by the citizens' advisory committee.

6.5
The Sequence of Events in a CRA

To allow the participants of a CRA to become acquainted and to prepare them for the upcoming work, a workshop is to be recommended. In most cases two days is sufficient for this. This introduces the participants to the CRA. Initially, the aims should be clarified to them. Furthermore, the participants are to be made familiar with the procedure of the CRA. A ranking exercise can be conducted for this purpose. In addition, accepted communication rules are to be adopted and the handling of possible conflicts explained during the workshop.

For the actual CRA process the next step is to arrive at an agreement about which problem areas one wishes the CRA to address. For instance, it is to be clarified for which regions and over what time span the CRA should be conducted. Furthermore, it is to be clarified which risk fields (transport-related risks, lifestyle risks, environment-related risks, etc.) should be considered. This requires a readiness to broaden one's perspective on problems, because a narrow view of problems obstructs the procedure. However, expansion ad infinitum is also problematic for the CRA.

This can be started with a brainstorming, which looks for risks in activity fields (housing and construction, communication, nutrition, mobility, etc.). What is to be striven for is a practicable list of problem fields to be treated. Within these fields, the risk categories (e.g., air pollution, waste, indoor air pollution, food contaminants) are chosen and then in turn target parameters, i.e., substances selected (e.g., acrylamide for foodstuffs or passive smoking for indoor air pollution, as well as particulate matter, O_3, NO_x, SO_2 for outdoor air pollution), which can serve as indicators for the risk categories. The appropriate attributes for these indicators, on the basis of the subjects of protection (health, environment, quality of life), are to be defined. These can be endpoints such as cancer disease or other attributes such as disturbances of wellbeing due to noise or nuisance odors. For these, metrics or risk metrics such as NOAELs, DALYs, or QALYs are selected from which a (risk) evaluation can be conducted. Which metrics and which methods of comparison are chosen depend on the data available and the aim of the comparison. Finally, comparisons are to be undertaken that will enable a ranking of risks to be made.

Depending upon which problem areas and which subjects of protection are considered, different activity areas are assigned to different expert teams to work on so that the process of risk assessment is not overly protracted. To this end, quite different approaches can be taken. For instance, working groups can be divided between environmental media (soil, air, water), between subjects of protection, or between fields of activity.[93]

Various methods of comparison can be used for the risk ranking. Broadly speaking, three variations are available. As well as the preferred multiattribute evaluation procedure, there are consensus building methods and simple voting procedures. In multiattribute procedures the participants of the citizens' advisory committee weight the evaluation of risks in various dimensions. From this, a ranking is arrived at. In contrast to this, risks in the other two procedures are placed directly into the ranking on the basis either of consensual decisions or the results of votes.

The instruments and methods used have an effect on the result of the CRA. They guide the attention, structure the problem field, and influence the ranking. We have reported the problems to be solved in this respect in Chapter 4.

Finally it must be determined how the process of the CRA and its findings should be documented, and which preliminary findings before the end of the process should be made available to the public, and in what form. What is for the most part not critical is the publication of successfully concluded procedural steps and the consensually reviewed preliminary findings.

A series of precautions need to be taken to ensure the success of a CRA. In this the main aims play a significant role. These aims, which above all affect the im-

---

[93] Experiences in the USA demonstrate that, especially in the division between environmental media, the danger exists that the experts (who work in authority in these areas) will be inclined to estimate "their" risks particularly high so as to keep their field of activity "high". Such interest-driven deviations can indeed never be entirely excluded but other divisions, for instance between fields of activity, can help to minimize this.

plementation of CRA findings and their broader effects in the sphere of environmental health policy, should be considered with care at the outset. Possible barriers such as institutional dogmas, thinking in pigeonholes, ideological conflict, and hidden interests are to be acknowledged and overcome. It would be naive to assume that this happens on its own.

6.6
The Organization of Communication

The nature of the problem should be grasped in the preparation of a CRA. Besides the factual level (which method should be used with which goal?), the interpersonal aspects play a significant role. One must here consider:
- What particular sensitivities do the involved individuals have?
- Is this the right time for a CRA?
- How far have the issues already escalated?
- Do all the parties have an interest in a CRA? If so, what?

For the participants, the collective clarification of the future working relations (e.g., through rules of procedure) as well as the agreement on the themes to be dealt with play an important role. A significant point in this is to achieve a consensus about the aim of the CRA and the roles of the participants. After that the procedure can be focused, the recommendations prepared, the options generated and assessed, or collective decisions be made. Whatever aims the participants pursue, they have different roles: as evaluators, advisors, or (co)decision-makers. Clear communication can help to avoid confusion over roles.

Every CRA requires that the participants are sufficiently well informed to be competent to contribute to a decision. For this it is necessary to address existing information deficits. All the necessary facts should be on the table. It may be required to organize hearings with experts so that all participants can form a competent opinion.

Tensions arise when different perceptions come up against each other. In such a situation it is necessary to give participants equal space for the demonstration of their standpoints, to explore the interests that lie behind their stances and at the same time to endeavor to determine and emphasize what the conflicting parties have in common.

CRA also requires that the risks to be compared are characterized to the best possible extent. The US Environmental Protection Agency (EPA) has discussed four criteria (see Table 25) by which the quality of a risk characterization may be appraised: transparency, clarity, consistency, and reasonableness (cf. US EPA 2000, pp. 14–19).

Transparency means the disclosure of the procedure of the risk assessment. This means that procedures, assumptions and models, and extrapolations must be described, gaps in the data identified, and uncertainties made clear. Clarity refers mainly to the comprehensibility of statements. Criteria here are conciseness and

Table 25 Criteria for risk characterization. (Source US EPA (2000, p. 19).)

Principle	Definition	Criteria for a good risk characterization
Transparency	Explicitness in the risk assessment process	Describe assessment approach, assumptions, extrapolations and use of models Describe plausible alternative assumptions Identify data gaps Distinguish science from policy Describe uncertainty Describe relative strength of assessment
Clarity	The assessment itself is free from obscure language and is easy to understand	Employ brevity Use plain English Avoid technical terms Use simple tables, graphics, and equations
Consistency	The conclusions of the risk assessment are characterized in harmony with other EPA actions	Follow statutes Follow Agency guidance Use Agency information systems Place assessment in context with similar risks Define level of effort Use review by peers
Reasonableness	The risk assessment is based on sound judgment	Use review by peers Use best available scientific information Use good judgment Use plausible alternatives

the logical structure of statements as well as comprehensible language. Consistence means the agreement with established procedures or guidelines for risk assessment. It must be made clear here whether the conclusions which are drawn from the findings are in agreement with the appropriate scientific standards and procedures. If this is not the case, it must be justified. Reasonableness, finally, is an overall criterion that indicates how sensible the evaluation of the risk is. The criteria here are that the risk characterization finds acceptance among the relevant scientific community and that it is based on the best data available and generally accepted scientific insights.

In any case, a series of faults and pitfalls are to be reckoned with here. Firstly, if risks are evaluated piecemeal, in various groups, then various assumptions, which have not been made explicit (e.g., the assumption of worst case or of reasonable case exposures), can lead to confusion (a lack of transparency). An overkill of risk jargon tends to hinder clarity. A particular problem for consistence is the application of different assumptions and procedures in risk assessments.

Besides such methodical and conceptual problems, there are also process-related aspects of communication to note. In principle these have to do with the implementation of fairness, competence, and trust (cf. Renn et al. 1995). These place high demands on communication.

Table 26 Causes of communication problems.

Step in CRA	Communication issue	Focus of communication
Preliminary phase	Distrust with respect to the organizers of the CRA	Disclosure and reporting of CRA-related objectives (subjects of protection and protection goals) Information about the approach to CRA and stakeholder participation in the CRA
Risk assessment	Dissent among experts	Information about and evaluation of the procedures used in hazard identification, dose–response assessment, and exposure assessment as well as characterization of the competencies and values of the experts
Risk characterization	Alleged intransparency and alleged confirmation biases in risk assessments	Transparent characterization of the applied procedures Understandable presentation of the most important results of the CRA
Risk ranking	Alleged nonconsideration of the societal value spectrum	Information about the CRA procedure and the involved stakeholders Open access to the data and results Opportunity to make comments

In parallel with internal communication, a sufficient external communication must be undertaken to ensure the acceptance of the CRA. Table 26 offers an overview of the communication problems encountered within the course of a CRA.

From a factual point of view, communication is particularly difficult if there are signs that a CRA is based on risk concepts and perspectives that deviate fundamentally from the public's risk perception. This problem can never be completely solved. A degree of tension always remains. If a CRA fully meets the public's risk perceptions, it fails to meet its aim and becomes a study in risk perception. However, a CRA that masks all aspects of public risk perceptions and touches only on scientific facts can go astray in society: the acceptance of its results will not take place. From this springs the task of explaining precisely the procedure of the CRA, and in particular the applied attributes.

Further risk communication problems with the public can arise from a lack of resources (time, personnel, finances), as well as a lack of consultation between the CRA project team and the CRA project management.

Whether a CRA is successful depends on how it is communicated to decisionmakers. Besides situational factors (time, circumstances) the presentation format plays a role. This above all means a good executive summary as well as graphic aids that summarize the important information and present it effectively. To date there are, however, no empirical investigations that address the question of the suitability of presentation formats for a CRA.

Overall, the experience with CRAs in the USA offers a rich source of material on the variety of implementation possibilities, but due to administrative, political, and not least cultural differences to Germany, or indeed Europe, this is of only limited transferability. If one wishes comparative risk evaluation to have a firm place in the process of risk regulation, practical trials of procedures such as those that have been laid out in this book are indispensable.

Appendix 1:
Risk Metrics

The choice of an appropriate risk metric is the task of experts in the relevant fields of risk. The availability of such a metric is a prerequisite for the appraisal of risks regarding their consequences for health. Various metrics have been proposed in the literature, of which the most significant are outlined below.

Metrics for characterizing dose–effect relationships:
NOAEL, benchmark dose, reference dose
The effect of a pollutant in an organism depends on the progression of the dose over time[94] and the way in which the dose is absorbed by the organism. When one speaks of a dose–effect relationship, then, a specific absorption path and a specific period of elapsed time is (implicitly or explicitly) assumed.

The relationship between dose and effect is assumed to be a continuous (mostly also a monotonically increasing) function. It varies, for example, with the age and general state of health of the individual, from individual to individual (intraspecies variability), and between various species (in particular between humans and animals, interspecies variability). In epidemiological experiments and laboratory tests only certain points of these functions can be determined, and these only with limited accuracy.

NOEL, NOAEL, LOEL, LOAEL, and benchmark dose are representative points of the dose–effect relationship:

NOAEL: no observed adverse effect level. The highest dose at which no adverse (morbid) effect can be established or at which no adverse effect is to be expected.

The specification of a NOAEL is not always possible. One of the following values may then be used as a substitute:

LOAEL: *lowest observed adverse effect level*. The lowest dose at which an adverse (morbid) effect can be established or at which an adverse effect is to be expected.

---

[94] For example, absorption of the entire dose, absorption of several individual doses, continuous absorption at a constant dose rate over a limited period, or continuous absorption at a constant dose rate over the entire lifetime.

Comparative Risk Assessment. Holger Schütz, Peter M. Wiedemann,
Wilfried Hennings, Johannes Mertens, and Martin Clauberg
Copyright © 2006 WILEY-VCH Verlag GmbH & Co. KGaA, Weinheim
ISBN 3-527-31667-1

NOEL: *no observed effect level.* The highest dose at which no effect can be established or at which no effect is to be expected.

LOEL: *lowest observed effect level.* The lowest dose at which an effect can be established or at which an effect is to be expected.

It is to be kept in mind when using LOELs and NOELs that these relate in general to biological effects. Since a CRA deals with possible harm to health, the use of LOELs and NOELs only makes sense when there are good grounds for assuming that the effects to which they refer can be taken as indicators for adverse effects.

It is to be noted when using a NOAEL that this does not prove with certainty that no adverse effect occurs at the respective doses under test conditions. This probability is, however, not explicitly stated and may be different for various substances due to differing test conditions. A more precisely defined value, the benchmark dose (BMD) is therefore proposed:

> Since the middle of the 1980s, the application of a Benchmark Dose, derived from the EPA's Reference Doses (RfD) has been discussed as a substitute for the NOAEL for substances with effect thresholds. In statistical terms, it represents the lower confidence belt of the effect dose ED_x, in which an effect occurs in x% of the test population examined.
> (Neus et al. 1998, pp. 175–176 [Translated from German])

An effect, which occurs in less than 10% or 5% of the test population, would be (depending on test conditions) on the border of what is identifiable, so that in fact the benchmark dose would lie in the region of the LOAEL or NOAEL (Barnes et al. 1995; cited in Neus et al. 1998, p. 176).

With all these threshold values, one must distinguish between the NOAEL (or benchmark dose) which the results of laboratory experiments with test animals show to be the highest dose at which no adverse effect can *be established,* and the NOAEL (or BMD) which, transposed to humans, represents the highest dose at which no adverse effect is to *be expected.*

The concept of the reference dose was developed by the US EPA to address various problems with the concepts of NOAEL and ADI (acceptable daily intake):

> The RfD [reference dose] is a benchmark dose operationally derived from the NOAEL by consistent application of generally order-of-magnitude uncertainty factors (UFs) that reflect various types of data sets used to estimate RfDs ... In addition, a modifying factor (MF), is sometimes used which is based on a professional judgment of the entire data base of the chemical...
> The RfD is determined by use of the following equation: RfD = NOAEL/(UF × MF)
> ...In general, the RfD is an estimate (with uncertainty spanning perhaps an order of magnitude) of a daily exposure to the human population (including sensitive subgroups) that is

likely to be without an appreciable risk of deleterious effects during a lifetime. The RfD is generally expressed in units of milligrams per kilogram of bodyweight per day (mg/kg/day).

The RfD is useful as a reference point from which to gauge the potential effects of the chemical at other doses. Usually, doses less than the RfD are not likely to be associated with adverse health risks, and are therefore less likely to be of regulatory concern. As the frequency and/or magnitude of the exposures exceeding the RfD increase, the probability of adverse effects in a human population increases. However, it should not be categorically concluded that all doses below the RfD are "acceptable" (or will be risk-free) and that all doses in excess of the RfD are "unacceptable" (or will result in adverse effects).

(US EPA 1993b)

All these metrics derived from the dose–effect relationship are threshold values and as such are not accumulable, i.e., they are not suitable for the appraisal of the accumulated effect of several pollutants.

Unit risk
Unit risk describes the estimated additional cancer risk to a human resulting from constant exposure over 70 years to a concentration of 1 µg of a hazardous substance per cubic meter of air (or equivalent concentrations in other media).

Becher et al. (1995, p. 88) define unit risk (UR) as the difference $UR = P_1 - P_0$. Here P_1 and P_0 are probabilities of death as a result of illness induced by the pollutant, integrated across the entire lifetime, P_0 with no burden of pollutant and P_1 with a burden of the pollutant at a concentration of 1 µg/m³ air (or equivalent concentration in other media). It is further stated that there are estimation procedures (by means of an approximation of the integrals through the sum) for ascertaining P_0 and P_1 that entail a (consistent) over- or underestimation of these values. Another estimation procedure, in which the observed additional risk R_x at a concentration x is divided by that concentration ($UR = R_x/x$), presupposes a linear dose–effect relationship. Thus only those unit risk values that have been calculated using the same estimation procedure should be compared.

What is to be noted here is that probabilities, and thus also unit risk values, are principally nonadditive. The arithmetical addition and subtraction of probabilities is an approximation which assumes that $P_0 < P_1 \ll 1$.[95] If unit risk values of several pollutants are added linearly, a risk value of >1 may result.

For the same reason, problems can arise if the illness or endpoint for which the unit risk should be determined constitutes more than a limited portion of all illnesses. In a borderline case for all illnesses (endpoints) together, $P_1 = P_0 = 1$. For a

[95] The exact formula for the probability of the combined risk is $P_1 = P_0 + UR - (P_0 \times UR)$; this reflects that fact that a person can die as a result of an exposure, or they can die as a result of some other thing, but they cannot first die of one and then of the other. Thus $UR = (P_1 - P_0)/(1 - P_0)$.

hypothetical pollutant which raises mortality rates for all illnesses, UR = 0 (Becher et al. 1995, p. 108).

The US EPA defines unit risk as follows (Kincaid et al. 1995):

> *Unit Risk: The upper-bound excess lifetime cancer risk estimated to result from continuous exposure to an agent at a concentration of 1 µg/L in water or 1 µg/m³ in air (with units of risk per µg/m³ air or risk per µg/L water).* (p. 5–38)
>
> *...The slope factor, or q_1^*, can also be used to determine the incremental cancer risk that would occur if the chemical was present in an environmental medium such as drinking water at a unit concentration (i.e., 1 µg of chemical per liter of drinking water). The calculation for drinking water usually assumes the person weighs 70 kg and drinks 2 liters of water per day:*
>
> *Drinking Water Unit Risk = $q_1^* \times 1/70$ kg $\times 2$ L/day $\times 10^{-3}$*
>
> *Air unit risk (risk per µg/m³) is derived from the linearized multistage procedure and calculated using the GLOBAL program.* (p. 5–46)

The slope factor referred to here is defined as follows:

> *Slope Factor (q_1^*): A measure of an individual's excess risk or increased likelihood of developing cancer if exposed to a chemical. It is determined from the upperbound of the slope of the dose–response curve in the low-dose region of the curve. More specifically, q_1^* is an approximation of the upper bound of the slope when using the linearized multistage procedure at low doses. The units of the slope factor are usually expressed as 1/(mg/kg-day) or (mg/kg-day)$^{-1}$.* (p. 5–37)
>
> *...The slope factor is a measure of the incremental risk or increased likelihood of an individual developing cancer if exposed to a unit dose of the chemical for a lifetime. The risk is expressed as a probability (i.e., one chance in ten or one chance in one million), and the unit dose is normally expressed as 1 mg of the chemical per unit body weight (kg) per day:*
>
> *Slope Factor = Risk per unit dose, or Risk per mg/kg-day*
>
> *When based on animal data, the slope factor is derived by extrapolating from the incidences of tumors occurring in animals receiving high doses of the chemical to low exposure levels expected for human contact in the environment. The EPA uses q_1^* for its risk assessments (see definition of slope factor). The q_1^* for a chemical, in units of (mg/kg-day)$^{-1}$, is based on the linearized multistage procedure for carcinogenesis and can be calculated by computer program (e.g., GLOBAL).* (p. 5–45)

The definitions of the US EPA also assume that the additional risk is <<1. Unit risk is also used by the US EPA to calculate accumulated risk arising from the burden of several carcinogens (US EPA Risk Assessment Forum Technical Panel 2000). This is also valid only as long as the resultant risk is <<1.

Relative risk (RR)
Relative risk is defined as the relationship of the risk of an adverse effect on health for an exposed person to the risk of an adverse effect on the health of an unexposed person (Kreienbrock and Schach 1995, p. 47ff). It indicates the extent to which the probability of an adverse effect on health rises when one undergoes a defined exposure. Relative risk can therefore be understood as a measure of the strength of the connection between exposure and adverse effect.

Individual exposure ratio (IER)
For noncarcinogenic effects, the EPA uses the IER, which is calculated as a quotient from the exposure dose and the reference dose (RfD). The IER indicates the extent to which the dose of a substance, which is absorbed, exceeds the safe dose. The higher the IER, the higher is the probability of adverse effects on health. Nonetheless, the EPA mentions a series of conceptual problems associated with the use of IER as a risk metric in CRAs (US EPA 1993a, p. 2.2–22). For instance, a risk comparison based on IER assumes that the same dose–effect relationship applies to all the substances under consideration – which is not the case.

Margin of exposure (MOE) and margin of safety (MOS)
According to Neus et al. (1998, p. 175) the margin of exposure is defined as follows:

> *Particularly in the context of regulation through authorization or licensing – as, e.g., in the EU's Technical Guidance Document (CEC 1996) – it is usual when appraising the health tolerance of an exposure to decide whether the margin of safety or margin of exposure is sufficient. This factor is, following the international agreement of the OECD and IPCS, defined as a quotient of the NOAEL (or another estimate for the effect threshold) and the estimated exposure; it can, according to this definition, only be given for substances with effect thresholds.*
> [Translated from German]

Margin of exposure and margin of safety are used as synonyms (cf. KEMI 2003, p. 35). It is to be noted that the term margin of safety is also used in another sense, that of a safety factor (e.g., Goldstein 1990).

The BfR (formerly the BgVV) uses the term margin of exposure as follows: "the MOE is calculated by dividing the dose which leads to tumors in animals by the quantity absorbed by humans" (BfR 2002). When comparing MOEs, therefore, attention should be paid to whether the effect threshold to which reference is made is the dose observed in animal experiments or that extrapolated form humans. This measure is not directly accumulable (for the appraisal of the simultaneous effect of several pollutants) but at best its reciprocal, and this also only for pollutants which lead via the same toxicity mechanism to the same endpoint (Sielken 2000).

Intake fraction

The intake fraction is a measure for the assessment of the transport of a pollutant between its release and exposed receptors (Bennett et al. 2002):

> *iF is the integrated incremental intake of a pollutant released from a source or source category (such as mobile sources, power plants, or refineries) and summed over all exposed individuals during a given exposure time, per unit of emitted pollutant.*

$$iF = \frac{\sum_{\text{people, time}} \text{mass intake of pollutant by an individual (mass)}}{\text{mass released into the environment (mass)}} \quad (A1)$$

> *Although the pollutant intake is summed over population and time, in actuality, when a pollutant is released, there is a distribution of individual exposures within the exposed population. Individual exposure can be quantified in terms of the individual intake fraction, iF_i. The total intake fraction comprises iF_is summed over all members of a potentially exposed population.*

This measure would only be accumulable if all pollutants were emitted in the same quantity, which is not usually the case.

Mortality, morbidity, and years of life lost (YLL)

Mortality designates the death rate in a population. When considering risks, one is primarily interested in cases of death attributable to a specific source of risk, which can be interpreted as a measure of the size of the risk. Mortality is used in this sense, for instance, in the comparative analysis of everyday risks (e.g., Cohen 1991; Wilson 1979).

Morbidity means the occurrence of disease in a population. When considering risks, one is primarily interested in the incidence of a disease, that is, the number of new cases of a disease. Again, what is chiefly of interest is the incidence attributable to a specific source of risk, for instance the number of cases of cancer which can be ascribed to exposure to a carcinogen.

Years of life lost (YLL), expected loss of lifetime (ELL), shows how much lower the life expectancy of an exposed person is than that of an unexposed person. In contrast to the pure counting of cases of death, YLL evaluates a case of death more highly the lower the age at which it occurs. The YLL metric can (as opposed to UR and MOE) also be applied to a number of end points, i.e., it is not disease-specific.

Disability adjusted life years (DALYs), quality adjusted life years (QALYs)
These metrics were developed so as to include in the comparison, along with various types of cases of death, diseases that do not directly lead to death.

QALYs are the integral of the state of health over a lifetime. A year of life in perfect health is counted with the value 1. Periods of illness are assessed with a weighting of <1. The circumstance of death is assigned the weighting 0. The weightings are arrived at by a survey of people. From this a weighting of <0 can result for diseases which are felt to be worse than death.

DALYs are the sum of years of life lost and years lived with a disability: DALYs = YLL + YLD. When calculating DALYs, years of life can be selectively discounted (with evaluation lower the further into the future) and weighted with a factor dependent on age.

According to Hofstetter and Hammitt (2002), there are few significant differences between QALYs and DALYs:
1. They are complementary, and changes to their values are inverse to each other (ΔQALY ~ $-\Delta$DALY).
2. Different methods are used to calculate the weightings for quality of life and disability.
3. DALYs take perfectly healthy people who die after the average expected lifetime as a reference. QALYs consider typical old-age disabilities and actual life expectancies.
4. DALYs can deal with an evaluation dependent on age, which cannot be applied with QALYs.

Two different points are to be noted here. Firstly, it may well be common practice to apply different methods to calculate the weightings for life quality and disabilities with DALYs and QALYS, though the definitions of DALYs and QALYs do not make this a necessity. One could also elect to use an evaluation dependent on age with QALYs. Secondly, if (as in point 2) different methods are used to calculate the weightings for life quality and disability, or (as in point 4) an age-dependent evaluation is taken for DALYs but not for QALYs, then point 1 (metrics are complementary, changes to values inverse to each other) is only approximately true.

DALYs and QALYs compare different states of health. These can, on the one hand, relate either to an individual or to a group of people for which the advantages and disadvantages of, for example, a medical treatment are to be weighed up; it is on this level that QALYs and DALYs are applied in evidence-based medicine. On the other hand, they can relate to one or more groups of individuals who are exposed to one or more pollutants. On this level DALYs or QALYS can be aggregated regarding the effects of various pollutants (e.g., US EPA 1998b). In the aggregation of groups, however, the risk is weighted less the fewer people belong to the group under consideration, e.g., a group of particularly sensitive people. Whether this weighting is "right", given the duty also to protect sensitive groups, is debatable (Hubbell 2002).

Problems with DALYs and QALYs:
- With QALYs both the severity and the length of an illness are taken as linear. This does not always agree with the actual appraisal of people affected.
- It is unclear what "perfect health" is. (The best state of health possible within a lifetime? The best possible state of health achievable for each age? Which group of people is to be taken as reference for "perfect health"?)
- It is debatable whether it should be discounted or not, and whether a valuation dependent on age should be made.

Willingness to pay (WTP)

WTP is an approach that converts health and death risks into financial values. WTP is the amount of money that the affected person is willing to pay to achieve an abatement of the risk. Willingness to accept (WTA) is the amount of money for which the affected person is willing to forego an abatement of the risk. As a rule, WTA > WTP, but the difference is sometimes slight. In practice, WTP and WTA are determined through interviews. (Further details regarding the distinction between WTP and WTA can be found in Hammitt 2002.)

WTP avoids the problem of the assumption of linearity that arises with QALY. However, with WTP it must be kept in mind that the ability to pay constrains the readiness to pay. Hubbell (2002) shows the differences in evaluation through QALY and WTP (Table 27).

Table 27 Comparison of QALY and WTP approaches. (After Hubbell (2002, p. 33).)

Parameter	QALY	WTP
Risk aversion	Risk neutral	Empirically determined
Relation of duration and quality	Independent	Empirically determined
Proportionality of duration/quality tradeoff	Constant	Variable
Treatment of time/age in utility function	Utility linear in time	Empirically determined
Preferences	Community	Individual
Source of preference data	Stated	Revealed and stated
Treatment of income and prices	Not explicitly considered	Constrains choices

The suitability of risk metrics for CRA

It is to be noted that when using these risk metrics in the context of CRA they relate to different dimensions. Some metrics deal with the dose–effect relationship; they characterize the toxicity of substances and thus their hazardous potential. Other metrics relate to exposure and still others to the extent of the ascertainable or expected effects on health. It is clear that metrics that relate to effect thresholds (NOAEL, etc.) cannot be straightforwardly compared with highly aggregated metrics such as QALYs and DALYs.

It is also to be noted that assessments for the same risk metric (e.g., risk to lifetime) but different endpoints (e.g., heart attack versus cancer) can be based on different methods that have a differing "sensitivity" to the appearance of risks (cf.

Neus et al. 1995, p. 305ff). So, for instance, the statistically assured limit of detection for lifetime-related mortality in environmental epidemiological experiments lies above 10^{-3} to 10^{-2}. In contrast, assessments for carcinogenic environmental burdens can be made which lie orders of magnitude (in the area of 10^{-5}) below this.

Finally one should notice that most risk metrics measure serious or even lethal effects on health. Until now, metrics with which the milder forms of adverse effects (e.g., disturbances to wellbeing due to noise or nuisance odors) can be measured and compared are lacking.

Which risk metrics can be used, under which conditions (and with which additional assumptions) can only be decided by the appropriate experts. With the participation of other stakeholders, it is important that what each individual metric expresses is made clear.

Appendix 2:
Multiattribute Comparative Risk Evaluation (MCRA)

The preparation and structuring of the evaluative process have a significant impact on the conduct of a multiattribute comparative risk evaluation (MCRA), particularly when stakeholder groups contribute alongside various experts. The profile of a multiattribute comparative risk evaluation presented in Table 28 follows from this.

The following two examples will illustrate how a multiattribute comparative risk evaluation can be conducted.

Table 28   Profile of a MCRA.

Aim	Comparison of various risks according to one or more attributes by one or more evaluators
Field of application	Known risks
Focus	Assessment of health risks
Participants	Experts and, in some cases, stakeholders
Output	Risk ranking, risk categorization

Example 1: MCRA procedure for environmental, health, and life-quality risks

The most comprehensive plan yet for a comparative evaluation of environmental, health, and life-quality risks is that laid out by the US EPA in their *Guidebook to Comparing Risks and Setting Environmental Priorities* (US EPA 1993a). It is based on the diverse experience that the EPA has gained from numerous national, regional, and local CRAs. These are discussed in detail in Chapter 3.

The EPA's approach (US EPA 1993a) is indeed fundamentally multidimensional; however, they do not follow these thoughts consistently to their end but allow the process to finish with separate rankings for various groups of effects. The reasons for this lie in part in a lack of data but mostly in the difficulties of finding a common risk metric for comparative risk evaluation (US EPA 1993a, p. 2.2-24).

It is our opinion that the approach to multiattribute evaluation presented in Section 5.2 offers the possibility of a complete and common risk evaluation and, thereby, of a unified ranking. However, the problem of lacking data cannot be overcome. As explained above, a multiattribute comparative risk evaluation is based on the data of risk assessments. Where this is missing, the multiattribute approach can also

Comparative Risk Assessment. Holger Schütz, Peter M. Wiedemann,
Wilfried Hennings, Johannes Mertens, and Martin Clauberg
Copyright © 2006 WILEY-VCH Verlag GmbH & Co. KGaA, Weinheim
ISBN 3-527-31667-1

help no further. One can attempt to arrive at a comparative risk evaluation with the help of qualitative appraisals. One such procedure is demonstrated below by the example of the WBGU appraisal.

Overview of the steps in a MCRA
The procedure of a MCRA can be divided into a series of steps with three phases. In the first, preparatory, phase the participants in the evaluative process are to be determined upon, i.e., expert teams and a citizens' advisory committee have to be selected. Secondly, the aims of the MCRA must be established. The following analytical phase is the domain of the experts. It comprises risk assessment and risk characterization. This last is the foundation for the final phase, the ranking of risks by the citizens' advisory committee. This procedure is, however, in no way linear – in particular, when citizens are involved. Then feedback takes place within the procedure, arising from the cooperation between experts and the citizens' advisory committee.

Step 1: Establishing the MCRA teams
Following a first examination of the problem fields, the expert teams and citizens' advisory committee are assembled. Both groups should be involved in decisions regarding the narrowing of the problem field and the choice of categories. It is above all the role of the experts at this stage to check the feasibility of the citizens' advisory committee's suggestions, because the data available play a significant role here and the choice of risk metrics depends decisively on the specification of categories. The result of this step in the process is a list of risks that should be considered in the MCRA.

Further, the experts have the task of determining more precisely the subjects of protection to be considered as well as the attributes and the risk metrics to be used for their measurement. They also deliver the risk assessments and the risk characterization, which is based on these in the form of data sheets, which are made available to the citizens' advisory committee for the comparative risk evaluation.

Step 2: Determining the cornerstones of the CRA
During the conduct of a CRA a series of decisions must be taken. These involve the creation of a problem list, the determination of attributes, and the choice of comparative procedure.

The problem list can be gained from, for instance, surveys, but can also be gained from a top-down procedure. In this, problem fields are selected for a region, e.g., from fields of activity (nutrition, etc.) and in the next step the problem fields (e.g., foodstuffs, drinking water), so as to arrive finally at the leading indicators (such as pesticide or acrylamide burdens). These must be chosen so that the significant risk problems in the problem field are covered by them. In this, difficult decisions have to be taken. It will have to be decided, in the sphere of indoor air, whether, besides radon and passive smoking, other indicators ought to be considered, such as formaldehyde. It is easy to appreciate that the choice of categories and indicators can substantially influence the results of a CRA.

So that the MCRA can remain clear, one limits the number of risks to be compared to, as a rule, a maximum of 20. This is above all the case when a larger number of attributes are to be used to make the comparison.

The choice of attributes follows the subjects of protection; in our case these are environmental health and perhaps quality of life. For the appraisal of environmental health risks the US EPA (1993a) here distinguished between carcinogenic and noncarcinogenic effects. For both types of effects, in principle at least, a risk metric is available. For carcinogenic effects this is the incidence of cancer, which is based on assessments of dose–effect relationship and exposure. For noncarcinogenic effects the EPA uses the individual exposure ratio (IER), which is calculated as a quotient of an exposure dose and the reference dose (RfD). The IER expresses the extent to which an absorbed dose exceeds the safe dose. The higher the IER is, the higher the probability of adverse effects on health. However, the EPA indicates a series of conceptual problems in the application of IER as a risk metric for CRA (US EPA 1993a, p. 2.2–22). For instance, a comparison of risks on the basis of IER presupposes that the same dose–effect relationship is valid for all the substances under consideration – which is not the case.

A central problem of this step is, therefore, the choice of appropriate risk metrics (see Appendix 1). The decision about which risk metrics are appropriate is one that in the end only experts can make.

As an alternative, the EPA suggests the use of risk metrics that relate directly to categories of harm. For carcinogenic risks the incidence of cancer is appropriate here. For noncarcinogenic risks, three categories of harm are suggested:

- *catastrophic effects on health*, e.g., death, reduction of life expectancy, serious physical or mental disability, hereditary diseases;
- *serious effects on health*, e.g., organ or nervous system dysfunction as well as behavioral or developmental disorders;
- *adverse effects on health*, e.g., loss of weight, enzyme changes, or reversible organic disorders.

The EPA discusses two further points which are relevant to the evaluation of health risks: uncertainties in the risk assessment arise above all from incomplete analyses, questionable data quality, and uncertainties in the assessment of exposure. An assessment of these uncertainties is at present only qualitatively possible in most cases, but can be considered as an independent attribute within the MCRA.

Sensitive subpopulations, meaning groups of people who react particularly sensitively to a specific pollutant, represent a particular problem for a MCRA. On the one hand, it is obvious that sensitive subpopulations must be considered within a comparative risk evaluation; on the other, the (not always only) scientific discussion regarding which groups are particularly sensitive to which substances is very controversial. Of special interest in the discussion is the question whether children are notably sensitive to environmental health hazards (e.g., Heinemeyer and Dürkop 2002; Landrigan 1999; Tamburlini et al. 2002). Sensitive groups can be treated as an independent attribute or attribute group – with a special weighting, if need be. If quantitative assessments are available, these can be used as described above for the normal population. Otherwise, estimations made with the help of rating scales must be resorted to here.

Risks to the quality of life should here be considered as a second aspect besides risks to health. In the usual way, economic effects are here taken into account, for instance health care costs, losses to agriculture, or damage to buildings. These are categories of damage for which as a rule quantitative data are on hand and which can be measured with a unified metric in the form of a currency unit (euro, dollar). More recent MCRA procedures in addition consider intangible aspects of quality of life, such as the loss of recreational areas or "aesthetic" burdens due to noise, loss of visibility in smog, or nuisance odors (US EPA 1993a, Ch 2.4).

The problem of measurement arises for these intangible aspects. Fundamentally, rating scales can also be used here. The question then occurs as to who should supply these ratings – in other words, who is an expert in the appraisal of these questions about quality of life? In contrast to health risks, which can only be assessed by experts, in the issue of loss of quality of life, every affected citizen is a potential expert. The problem here is to obtain evaluators that are representative of those affected. In the simplest case, a citizens' committee can be resorted to, if this can be representative of loss of quality of life for the affected population. Otherwise, representative surveys must be conducted to acquire the necessary data. The issue of representative appraisals also arises for another procedure for the investigation of appraisals of intangible aspects: ascertaining willingness to pay.

The final task in this step is the selection of a comparative procedure. In principle, the procedure elucidated in Section 5.2 is available for this. Which procedure is preferred depends on the complexity of the problem and the time available, as well as on the quality of the decision-making demanded. The multiattribute evaluative procedure discussed in Section 5.2 is certainly the most laborious, but offers the highest degree of transparency and rationality in the evaluation.

Step 3: Risk assessment
The conduct of risk assessments is the domain of experts. There are by now numerous papers on their methodology, the discussion of which would go far beyond the scope of this book (on this point see also Section 2.4). What is particularly interesting for the conduct of CRAs are the more recent approaches which aim to harmonize diverse practical procedures (EC 2000a,b).

From the cornerstones established in step 2 above, a number of requirements arise relating to the risk assessment. For instance, it must be ascertained for which of the problem fields or indicators selected risk assessments already exist, and on which risk metrics these are based.

Should no assessments exist for the space and time range selected, one can seek to transfer assessments from similarly situated cases or circumstances. The same is true for attributes relating to sensitive subpopulations. The unavoidable uncertainties attendant on such assessments must then be considered in the comparative assessment. The explicit explanation of the uncertainties associated with risk assessments is an elementary component of the communication of assessment results.[96]

Step 4: Risk characterization
The purpose of risk characterization in the context of MCRA is a transparent and expressive presentation of the findings of the risk assessments. One possibility is to summarize the assessment findings for the various attributes of each risk on a separate data sheet. Figure 20 shows an example for the risk of a school bus accident from a comparative risk evaluation for pupils of a US school (Florig et al. 2001). The data sheet contains a brief textual summary of the risk assessment and a table with the quantitative assessments for the individual attributes. In this MCRA various metrics were supplied for the attributes of mortality and morbidity (e.g., frequencies and probabilities). This is intended to provide a better understanding of the risk. In the context of a multiattribute evaluation, however, one of these risk metrics must be selected for each attribute (mortality and morbidity), so that this metric is compatible with the risk characterizations of other risks.

The data sheet in Figure 20 also provides an example of how the uncertainties associated with risk assessments may be presented. Not only the best estimate, but also the values for the highest and lowest assessments are given. Other possibilities for communicating uncertainties are to indicate confidence intervals or percentages. Naturally, one must ensure that those taking part in the MCRA understand the terms in which uncertainties are stated.

In practice risk characterizations are mostly more extensive and complex than is the case in this simple example from Florig et al. (2001). Even so, it should be possible to produce risk characterizations for individual risks in the form of summary data sheets, which provide participants with a clear evaluation. Examples of such comprehensive data sheets are to be found in the New Jersey Comparative Risk project (NJDEP 2003, Appendix 4).

Step 5: Ranking
The final step of ranking can be configured variously, depending on the comparative procedure chosen. In principle, three variations are practicable:
- The participants evaluate all the risks together on the basis of the data sheets and then form the risks into a ranking following their holistic judgment. Procedurally, this is probably the simplest approach. Moreover, it corresponds with the everyday intuitive method of judgment. This procedure therefore is suggested if the comparative evaluation is to be performed by a citizens' advisory committee. However, this holistic approach quickly reaches the limits of the human ability to handle information when a large number of risks and attributes are to be evaluated.

---

[96] The importance of the consideration of uncertainties associated with risk assessment is emphasized in numerous works on the problems of risk assessments: see, e.g., Bailar and Bailer (1999), US EPA (1996), National Research Council (1994).

School Bus Accidents

Summary:

Most school bus-related deaths occur among students who are outside the bus either getting on or getting off. Half of school bus injuries occur among students on the bus. At Centerville Middle School half of the 430 students ride the school, almost identical to the national average. Accidents involving more than one death are very rare. Because CMS buses use the Alvarez Expressway and cross the C&LL rail line, the risk of a catastrophic bus accident in Centerville is estimated to be between four and six times higher than the national average.

School bus accident risk for Centerville Middle School

Student deaths	Low estim.	Best estimate	High estim.
Number of deaths per year	.0001	.0002	.0004
Chance in a million of death per year for the average student	.25	0.5	1
Chance in a million of death per year for the student at highest risk	0.5	1	2
Greatest number of deaths in a single episode		20-50	

Student illness or injury			
More serious long-term cases per year	.0002	.0006	.002
Less serious long-term cases per year	.0004	.0015	.004
More serious short-term cases per year	.001	.002	.006
Less serious short-term cases per year	.002	.005	.015

Other Factors	
Time between exposure and health effects	**immediate**
Quality of scientific understanding	**high**
Combined uncertainty in death, illness, injury	**1.6 (low)**
Ability of student/parent to control exposure	**moderate**

Figure 20
Example of a data sheet for risk characterization in a MCRA.
(From Florig et al. (2001, p. 919).)

- The evaluation and the ranking proceed by a noncompensatory ordering rule, for instance the lexicographic rule.[97] Following this, each participant (or participant group) firstly selects the most important attribute and takes the risk assessment of this attribute from the data sheet of each risk. A ranking is formed from these assessments. Should two or more risks have the same assessments for this attribute, and thus occupy the same place in the ranking, then the evaluation can proceed following the second most important attributes, and so on. While this procedure remains within the bounds of cognitive effort, it is not very transparent, particularly when extended to different rankings.
- The comparative risk evaluation proceeds with the help of a multiattribute evaluation procedure, as explained in Section 5.2. The conduct of such a procedure enables a transparent structuring of the problem in which differences in evaluations and their influence on the resultant risk ranking, and thus in the evaluation process, can be identified. While naturally this does not lead automatically to the elimination of such differences, it is a substantial prerequisite for a rational discussion.

In the following, the aspects of attribute structuring and weighting and the problems arising from them are illustrated using the EPA example introduced above.

The US EPA (1993a) distinguishes between carcinogenic and noncarcinogenic risks. This leads to an attribute structure such as that presented in Figure 21a. Here it should be noted that the three terminal attributes together can at maximum only be as important as their superordinate "noncarcinogenic" attribute. If one weights the two superior attributes equally strongly, that means that the terminal attribute of "incidence of cancer" carries 50% of the total weight of all attributes of health risks, and the three attributes of the aspect of "noncarcinogenic" must then share the remaining 50%. The weights accorded to each attribute level are included in the figure.

This effect becomes even stronger when one weights carcinogenic risks more highly than noncarcinogenic – which is probably a frequently encountered intuitive evaluation. The weighting can proceed more directly and understandably if the attributes are structured nonhierarchically, as depicted in Figure 21b. Here all four attributes can be directly compared with, and weighted against, each other. The values arising here from a simple rank weighting exhibit less sharp differences.

A MCRA for such attributes can be worked through as an example for three risks using data contained in the EPA's guidebook (US EPA 1993a, p. 2.2–31). The upper block of Table 29 contains assessments of the numbers of people endangered due to air pollutants, drinking water pollutants, and hazardous waste. The risk ranking provided by the raw data is unambiguous: the health risks from air pollutants are far higher than those from drinking water pollutants and hazardous waste. This ranking also does not change when the data are standardized (middle block in Table 29) and weighted (lower block in Table 29) following the procedure outlined in Section 5.2.2. This is also an example of a marked dominance structure among

---

97) See Section 5.2.2.

172 | Appendix 2: Multiattribute Comparative Risk Evaluation (MCRA)

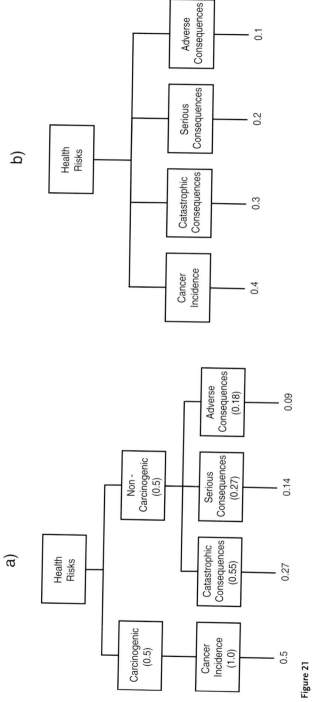

Figure 21
Two possibilities for structuring attributes.

Table 29 Example of MRCA on the basis of EPA data.

Risks	Incidence of cancer	Catastrophic consequences	Serious consequences	Adverse consequences	Σ
Air pollutants	12.5	5000	250,000	800,000	
Drinking water	4.0	2000	50,000	250,000	
Hazardous waste	1.2	100	5000	10,000	
↓					
Air pollutants	1.000	1.000	1.000	1.000	
Drinking water	0.248	0.388	0.184	0.304	
Hazardous waste	0.000	0.000	0.000	0.000	
↓					
Weight	0.400	0.300	0.200	0.100	
Air pollutants	0.400	0.300	0.200	0.100	1.000
Drinking water	0.099	0.116	0.037	0.030	0.283
Hazardous waste	0.000	0.000	0.000	0.000	0.000

the risks. Air pollutants have a higher value on all attributes than those for drinking water, and these have again higher values than all those for hazardous waste. In such a case, standardization and weighting are pointless, because the ranking arises necessarily from the raw data.

As the complexity of the attribute structure increases, so does the significance of a procedure for the application of attribute weights. Usually, one will resort to a procedure in which the individual subgroups are separately weighted and then the absolute weights are calculated by multiplying through the branches (see Section 5.2.2).

Example 2: Procedure of a MCRA for very heterogeneous risks

Particularly when comparing very heterogeneous risks with differing harmful consequences and different time and spatial horizons, a quantitative assessment of harmful consequences is no longer possible. Here one must resort to a qualitative evaluation. These can, however, be quantified with the help of rating scales.

This procedure will be illustrated by the example of the WGBU (1998) annual report, *Welt im Wandel – Strategien zur Bewältigung globaler Umweltrisiken*.[98] The aim of the report was not to create a risk ranking, but through this characterization establish various risk management strategies. Nonetheless the analysis offers a good example of the problems of comparative risk assessment when the evaluation of risks is connected with high uncertainties and when quantitative assessments are not possible.

In the WGBU report a very heterogeneous list of (potential) environmental risks are evaluated using various attributes. These attributes relate on the one hand to the possible damage (differing both in the extent of the damage and the certainty

---

[98] See also Chapter 3.

Appendix 2: Multiattribute Comparative Risk Evaluation (MCRA)

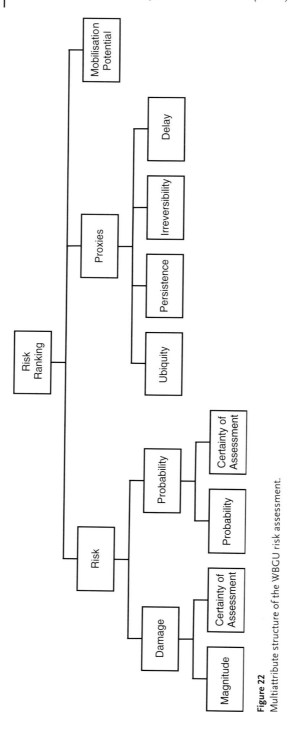

Figure 22
Multiattribute structure of the WBGU risk assessment.

of the assessment of the extent of the damage) and the probability (probability of occurrence and certainty of assessment of probability of occurrence). On the other hand, assessment aspects were used as "proxies" that characterize the risks regarding time and spatial aspects. These are ubiquity, persistence, irreversibility, and delay. One further criterion is the potential for public mobilization, relating to the assessment of the potential for social conflict arising from the risk. The WBGU founded this selection using the results of risk perception research, in which these aspects have been demonstrated to be significant for the intuitive appraisal of risks. Figure 22 shows these criteria in a hierarchical attribute structure.

Using these attributes, the WBGU evaluated a list of more than 20 (potential) risks, from technological risks (e.g., nuclear power, large chemical plants) through health risks (e.g., AIDS infection, BSE–v-CJD infection), environmental risks (release of genetically modified plants), and pollutant risks (e.g., persistent organic pollutants, endocrine disruptors) to climate risks (e.g., gradual anthropogenic climate change, instability of the West Antarctic ice sheet) and risks arising from natural events (e.g., flooding, earthquakes).

The evaluation of these risks according to the attributes proceeded with the help of four-level rating scales with verbal labels ("low", "rather low", "rather high", "high", as well as a separate "unknown" category), with which the evaluation could also comprise several grades (something represented by shaded fields on the scale; see Figure 23). The evaluations of the risks themselves were based on comprehensive problem analyses, which, however, did not lead to results that would allow a quantitative assessment according to the attributes. The high degree of uncertainty associated with practically all the risks considered was explained by the fact that the WBGU in this report concerned itself exclusively with risks that did not fall into the category of normal risks and were therefore marked by high uncertainty.[99]

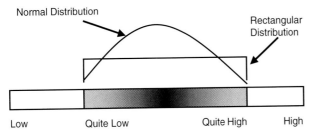

Figure 23
Example of possible interpretations of unclear assessment on the WGBU (1998) rating scale.

[99] Normal risks are distinguished thus: low uncertainties associated with the probability distribution of damage, overall a rather low potential for damage, overall low to medium probability of occurrence, low persistence and ubiquity (time and spatial spread), extensive reversibility of potential damage, low margin of deviation of potential damage and probability of occurrence and low potential for mobilization and (social) conflict (i.e., no significant discrepancy of evaluation between the group of risk carriers and the group of chance or benefit winners) (WBGU 1998, p. 6ff).

The WBGU accommodated the aspect of uncertainty through the two aspects of "certainty of the assessment of the extent of the damage" and "certainty of assessment of probability of occurrence". Such a conceptualization of uncertainties may well be of help for the purpose of classifying risks so as to formulate risk management strategies – which is why it appears in the WBGU report. In the context of a multiattribute evaluation, though, the fundamental problem arises that it is not clear what quality a high or low certainty of assessment has. Should a low certainty of assessment be taken as evidence for or against the existence of a risk?

In regards to the probability of occurrence, the certainty of assessment can be understood as an assessment of probability of the second order. Depending on the form given to this probability distribution (e.g., rectangular or normal distribution; see Figure 23), different interpretative possibilities arise: with the rectangular distribution, for instance, all the values within the interval under consideration are equally probable; separate analyses are advisable for the upper and lower borders of the distribution. Assuming an approximately normal distribution (indicated in Figure 23 by the shaded area), one can use a scale value for the analysis which corresponds to the average.[100]

Thus the criterion of certainty of assessment can be replaced by probability of occurrence through a separate analysis of upper and lower interval limits or through the average for a multiattribute structuring using WBGU criteria. This gives the attribute structure shown in Figure 24.

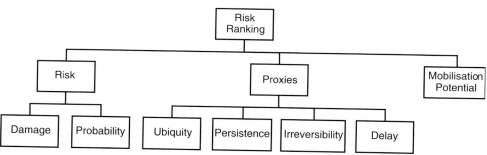

Figure 24
New attribute structure of WBGU risk assessment.

In this structure, damage and probability form the two attributes to be appraised for the superordinate attribute of risk. It should be noted that this conceptualization of risk differs fundamentally from that usual in risk assessment (see Chapter 2). Here damage and probability are additively linked. Thus a higher damage cannot be compensated for by a correspondingly low probability (or vice versa). This appears, however, from the point of view of the qualitative appraisal of damage and probability to be appropriate. Basically, it would also be possible here to use risk as a product term, the more so since WBGU itself defines the usual probability inter-

---

[100] The shading indicates an uneven distribution so that the scale value, which corresponds to the median, can be used.

Appendix 2: Multiattribute Comparative Risk Evaluation (MCRA)

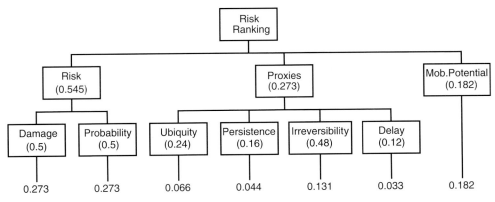

Figure 25
Attribute structure of WBGU risk assessment with absolute weights (relative weights in parentheses).

val (0:1) as the underlying scale for its assessments of occurrence probabilities.[101] However, it is difficult to rate both the extremely low probabilities, which have to be assumed for some risks (e.g., nuclear power, meteorites),[102] and the rather high probabilities for other risks on this scale (see Table 30).

In the following an example comparison based on multiattribute analysis is made using some of the more than 20 risks analyzed by WBGU. Figure 25 shows the attribute structure once more, but this time with example weightings. These weightings were (to simplify matters) made following the rank weighting procedure (see Section 5.2.2). The attribute "risk" was allotted rank 1, "proxies" rank 2, and "potential for public mobilization" rank 3. Damage and probability were equally weighted and the ranking series irreversibility, ubiquity, persistence, and delay was assigned to the four proxy attributes. In the lower row of Figure 25 the absolute weights – used later in the calculation – are shown. What is interesting is that, with this specific structure and the weighting procedure used, the attribute of potential for public mobilization obtains the third highest weight, although (at the highest attribute level) the attribute "proxies" has a significantly higher weight. It was demonstrated in Section 5.2.2 that one must check whether such effects of hierarchical attribute structuring agree with the actual estimations of attribute weights from the assessors.

[101] WBGU (1998, p. 60) explains its scale of probability of occurrence thus: "low means: 'very improbable' (near 0)"; "high means 'very probable' (near 1)".

[102] The probability of a catastrophic nuclear accident can be determined to be between 10^{-6} and 10^{-5} per year of operation (NEA 2000, p. 38). A (very high) estimate for the probability of a global catastrophe through meteor strike with up to a billion dead amounts to four strikes per million years (Gerrard 2000, p. 901). At an OECD workshop (OECD 2003) the probability of a catastrophic asteroid strike with the potential to destroy an entire continent in the twenty-first century was estimated at 1:5000. The probability of a globally catastrophic asteroid strike in the twenty-first century was estimated as 10^{-6}.

For the example risk comparison the following risks were selected from the WBGU list: destabilization of the ecosystem, electromagnetic fields, nuclear power, meteorites, (natural) climate variation, (anthropogenic) climate change, and floods. In the WBGU report every risk has appraisals for the attributes. These appraisals are given in the form of (more or less unambiguous) marked fields on a four-level rating scale. For the multiattribute assessment, the label of the rating scale was assigned a numerical value: "low" = 1, "rather low" = 2, "rather high" = 3, and "high" = 4. With ambiguous markings which spanned more than one scale level, an average was estimated (see example above in Figure 23). The upper block in Table 30 shows the appraisals so extrapolated: the sum column on the far right contains the line sum (and thus gives a first indication of the ranking of the risks).

In the next step these raw values were – as outlined in Section 5.2.2 – standardized on a risk index with the interval (0:1). The middle block in Table 30 shows in the upper line the weight, as given in Figure 25, and under this the (as yet unweighted) standardized values of the risk index. In the last step, these standardized values were multiplied by the weights. The final column contains the information important for the comparison and ranking. It contains the sum of the weighted individual appraisals of the attributes.

The findings of this multiattribute evaluation process are given once more in Figure 26. It shows that, for this selection of risks and using WBGU appraisals, the destabilization of ecosystems, anthropogenic climate change, and nuclear power are clearly the greatest risks. What is interesting is that meteorites and electromagnetic fields have risks of nearly the same magnitude. This case also illustrates the significance of sensitivity analyses. A glance at Table 30 shows that meteorites

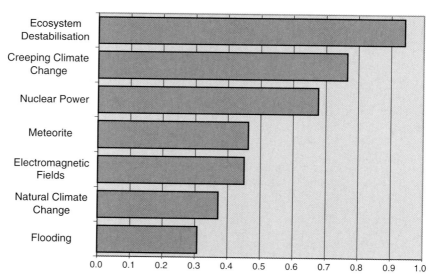

Figure 26
Ranking of environmental risks on the basis of a multiattribute risk index.

Appendix 2: Multiattribute Comparative Risk Evaluation (MCRA)

Table 30 Example of multiattribute risk assessment on the basis of WBGU data.

Source of risk	Probability of occurrence	Extent of damage	Ubiquity	Persistence	Irreversibility	Delay	Potential for public mobilization	Σ
Dest. ecosystem	3.500	4.000	4.000	3.700	3.700	4.000	3.000	25.900
Electromagnetic fields	2.000	2.000	3.500	1.500	3.000	2.000	4.000	18.000
Nuclear power	1.000	4.000	3.700	3.700	3.700	1.700	3.700	21.500
Meteorites	1.200	3.600	3.600	3.300	3.700	1.000	1.000	17.400
Nat. climate variation	2.000	3.000	2.500	1.300	2.000	1.300	2.000	14.100
Climate change	3.000	3.700	3.600	3.700	3.600	3.700	2.000	23.300
Floods	1.300	3.600	1.700	3.500	1.300	1.300	1.200	13.900
Weight	0.273	0.273	0.066	0.044 →	0.131	0.033	0.182	1.000
Dest. ecosystem	1.000	1.000	1.000	1.000	1.000	1.000	0.667	6.667
Electromagnetic fields	0.400	0.000	0.783	0.083	0.708	0.333	1.000	3.308
Nuclear power	0.000	1.000	0.870	1.000	1.000	0.233	0.900	5.003
Meteorites	0.080	0.800	0.826	0.833	1.000	0.000	0.000	3.539
Nat. climate variation	0.400	0.500	0.348	0.000	0.292	0.100	0.333	1.973
Climate change	0.800	0.850	0.826	1.000	0.958	0.900	0.333	5.668
Floods	0.120	0.800	0.000	0.917	0.000	0.100	0.067	2.003
Dest. ecosystem	0.273	0.273	0.066	0.044 →	0.131	0.033	0.121	0.941
Electromagnetic fields	0.109	0.000	0.052	0.004	0.093	0.011	0.182	0.450
Nuclear power	0.000	0.273	0.057	0.044	0.131	0.008	0.164	0.677
Meteorites	0.022	0.218	0.055	0.037	0.131	0.000	0.000	0.462
Nat. climate variation	0.109	0.137	0.023	0.000	0.038	0.003	0.061	0.371
Climate change	0.218	0.232	0.055	0.044	0.126	0.030	0.061	0.765
Floods	0.033	0.218	0.000	0.040	0.000	0.003	0.012	0.307

and electromagnetic fields vary most for the attribute "potential for public mobilization". As discussed above, this attribute received a relatively high weighting due to the attribute structure. A sensitivity analysis in which the weighting for "potential for public mobilization" is systematically reduced and those of the two other attributes "risk" and "proxies" proportionally increased, shows that in fact electromagnetic fields and meteorites exchange their ranks when the weight of "potential for public mobilization" is approximately halved. However, a stronger reduction in the weight of "potential for public mobilization" leads to no further changes in the ranking.

Appendix 3:
Comparative Evaluation of Unclear Risks

In the evaluation of unclear risks (EUR), the essential problem is that various evaluators can arrive at different outcomes even in the appraisal of scientific findings. For this reason, this circumstance should occupy the central point in the design of the EUR. The purpose of EUR is the comparison of evaluations of an unclear risk by various evaluators with the aim of forming a consensus regarding the evidence for the existence of a hazard (see Table 31). The transfer to the comparative evaluation of several unclear risks occurs relatively simply.

Table 31 Profile of an EUR.

Aim	Determination of a hazard; determination of dissent/consensus between various evaluators in the identification of a hazard
Field of application	Unclear risks
Focus	Evidence of a hazard
Participation	Experts with various hazard evaluations, citizens' advisory council
Output	Evidence evaluations and evidence rankings

The US EPA, the Australian National Health and Medical Research Council (NHMRC), and the Cochrane Collaboration, as well as the International Agency for Research on Cancer (IARC), the OECD, and the EU have developed a series of standards on which an EUR may be built. These deal with guidelines for literature searches, analysis and presentation of scientific experiments, hazard identification, and hazard characterization.[103] We follow these guidelines in our proposals for EUR.

In the following the EUR of the hazards associated with high-frequency electromagnetic fields is presented as an example, because it deals with a prototypical case that has been controversially discussed and has led to repeated conflicts in recent years. Full particulars are to be found in the DIFU and WIK reports (Drüke et al. 2003; Büllingen et al. 2002; but see also Wiedemann et al. 2002a).

[103] See, among others, ICNIRP (2002), National Health and Medical Research Council (1999, 2000), Presidential/Congressional Commission (1997), National Research Council (1996), EPA Risk Assessment Guidelines (http://www.epa.gov/ncea/raf/rafguid.htm), technical guidance document.

Comparative Risk Assessment. Holger Schütz, Peter M. Wiedemann,
Wilfried Hennings, Johannes Mertens, and Martin Clauberg
Copyright © 2006 WILEY-VCH Verlag GmbH & Co. KGaA, Weinheim
ISBN 3-527-31667-1

Appendix 3: Comparative Evaluation of Unclear Risks

Overview of the EUR's steps

As Figure 27 shows, a series of tasks are here to be completed. Firstly, the aim of the EUR has to be specified. Then the EUR team, comprising experts and a citizens' advisory council, has to be formed. Next, the experts begin work. Their first task is the designation of the data on which their hazard assessments are to be based. A further step is the collection and assessment of the studies to their various endpoints. After this the hazard characterization proceeds in the form of arguments for and against, on which the evaluation of the citizens' advisory council is formed.

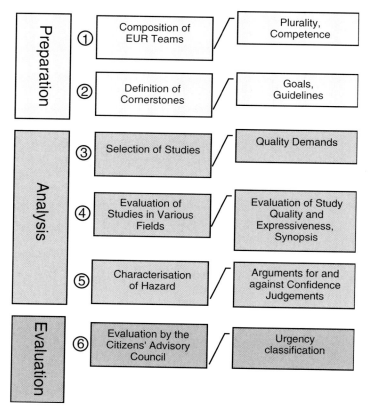

Figure 27
Steps of an EUR.

Step 1: Designation of the EUR team

The starting point of an EUR is, as a rule, dissent over the existence of a risk, i.e., unclear risks take center stage.

As discussed in Chapter 6, expert teams and a citizens' advisory council are to be established. The job of the expert teams is the assessment of evidence for the supposition of a risk. The job of the citizens' advisory council is the weighting of arguments for and against the hazard or a setting of priorities.

Since the citizens' advisory council can only act after the experts' work is done, the selection of experts is dealt with first.

The composition of the expert team is of essential significance to the EUR. On the one hand, the process deals with the differing appraisals of the hazard on the part of experts. On the other hand, complex questions of detail are to be answered which require the input of knowledgeable experts in the individual subject areas.

We therefore propose that experts representing the poles of dissent be included, and that additionally such experts as have specialist competences in a subject area or experimental sphere, but, wherever possible, have taken no part in previous controversy over the existence of the hazard, should participate (see Figure 28).

The reasons, in detail, are:
- The findings of research from numerous and, to some extent very different, subject areas must be evaluated for the hazard evaluation. Because the evaluation demands quite different competences, care must be taken that as many of the most significant areas as possible are represented.
- Besides the obvious criterion of competence in an area of expertise, two further criteria for the choice of advising experts are to be noted: they should be independent of any commercial or political interest in the debate about risk. It should thus be ensured that the advising experts may not be accused of any partisanship which, in the public discussion of a highly politicized issue, can be laid open, for instance to discredit disagreeable appraisals.

To improve the transparency of the expert knowledge within the group, competence profiles should be prepared and made publicly accessible. This means that every expert has to declare his or her scientific core competences, which fields of activity he or she represents, and to which scientific bodies he or she belongs. The Cochrane Collaboration (2003) presents a similar argument.

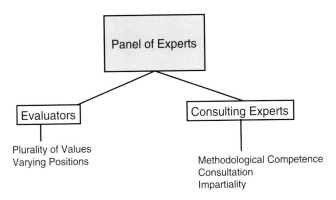

Figure 28
Composition of the expert group.

Proposals about the composition of the expert group should also be the subject of consensus so as to retain the ability of the group to operate.

The citizens' advisory committee should represent various societal stakeholders, e.g., local authorities, associations, industry, and citizens. It can also be arranged after the model of the planning cell and someone from a selected group of nonorganized citizens included at random.

The citizens' advisory committee should be included in the evaluation of evidence. It is their task to evaluate the arguments for and against (see step 6 below).

Step 2: Establishing the cornerstones of the EUR
What is essential is a precisely defined goal. It is indeed clear that it deals with the appraisal of evidence for an unclear risk. What speaks for the existence of a risk, and what against? The difficulty, as ever, lies in the details. The first review necessary regards the data pool. Scientific investigations, the professional knowledge of doctors about individual cases, and the knowledge of lay people are all available for selection. What knowledge shall be included, and what excluded? And if all three classes of knowledge are considered, how shall decisions be made in the case of disagreement?

In the analysis of a hazard, scientific knowledge should be granted the decisive role. This does not mean that the other classes of knowledge should have no significance. Professional and lay knowledge should be brought in to the discussion of a suspicion of risk. However, because of the known possibilities of error in the intuitive judgment process (see Nisbett and Ross 1980) they are not appropriate for the testing of a risk hypothesis.

A second review must be performed regarding the goals of the assessments of hazard. The discussion about the precautionary principle plays a role in this. What must be clarified are the central concepts with which the body of evidence shall be characterized. Agreement must be reached about the operational definitions for such concepts as proof, suspicion, indication, adversity, etc., to be used. Wherever possible, the definitions introduced by recognized bodies such as the WHO should be resorted to.

A further question is whether additional attributes for the hazard analysis should be selected besides the health-related endpoints. This in particular concerns the potential for public mobilization (see WBGU 1998), a term which refers to risk perception and the mechanisms of social reinforcement. We remain skeptical about the inclusion of such an attribute, since it seems doubtful that this can make a contribution to hazard analysis. The same goes for the WBGU criteria of ubiquity, irreversibility, delay, and persistence. However, should a hazard identification not succeed, these can be consulted as additional criteria for precautionary management.

Step 3: The selection of relevant studies
The evaluators of the EUR need to reach agreement on which studies should be included in the evaluation of effects on health. Preliminary steps towards this have already been taken in step 1 (establishing the aims). However, a unified conception

for the selection and evaluation of the scientific literature and the creation of the overall scientific picture must be developed.

The procedure employed in the literature search is also relevant. In this the search strategy for the identification of meta-analyses in Cinahl and Medline[104] can be resorted to; more can be found on this topic in the Cochrane Reviewers Handbook. This is because different evaluators, despite working under the same conditions, may differ considerably in the compilation of the literature (cf. Wiedemann et al. 2002a).

The Cochrane Reviewers Handbook provides very detailed information on all facets of a systematic search and evaluation of literature, especially on randomized controlled studies, for the selection and evaluation of scientific literature.[105] The Handbook of the British NHS Centre for Reviews and Dissemination is similarly detailed and extensive.[106] Determining the quality of scientific studies is salient in their selection. Central questions are:
- How reliable is the exposure assessment?
- What requirements were set on the specificity and sensitivity of the procedures used?
- What is considered as a health-relevant effect (the question of adversity)?
- How is heterogeneity dealt with?

Furthermore, agreement must be reached on what record should be made of the compilation of the relevant studies, and which criteria should be applied in the evaluation of the studies' quality (see Table 32).

Table 32 Examples of criteria for the quality evaluation of a study to be included.

	High weighting	Low weighting
Animal experiments	Exposure reliably assessed	Exposure measurement rather defective
	Low rates	High incidence of spontaneous tumors
	High specificity of experimental procedure	Low specificity of experimental procedure
	Dose–effect relationship	No data
Human data	Exposure reliably assessed	Exposure measurement rather defective
	High specificity of investigative procedure	Low specificity of investigative procedure
	Control of biases and confounders	Low or no control

[104] http://www.york.ac.uk/inst/crd/search.htm
[105] Clarke and Oxman (2002).
[106] NHS Centre for Reviews and Dissemination (2001).

The record must also state which studies have been excluded, and on what grounds.

Step 4: Examining the scientific experiments in the various problem areas
Following agreement about the data to be used, what the experiments express is then examined. A reference method that may be used is that of evidence-based medicine, which provides rules on how to proceed from the examination of scientific experiments to the construction of an overall scientific picture.[107] For the topic of electromagnetic fields, this means the evaluation of existing experiments according to the criteria laid out in Table 33.

A summary, which forms the basis on which the arguments for and against are later formulated, is to be created for each problem area. The evaluators have to agree on a form of presentation, which should follow the criteria in Table 33.

[107] Cf. National Health and Medical Research Council (2000).

Table 33 Examples of criteria for the evaluation of studies.

Criterion	Commentary
What the experimental approach expresses (general suitability for the examination of a causal connection)	Experiments can give falsely positive or falsely negative results. They should therefore be critically examined. The principal suitability of the experiments is to be appraised. Is the error rate of the tests used in the studies tolerable?
Security of findings against coincidence, interfering variables, and errors of measurement	The results of experimental approaches, which are in principle suitable, also need to be critically examined. One needs to decide whether the findings have not also been produced by coincidence. Have random samples and other errors been minimized? Have interfering variables been excluded or at least controlled for?
Reliability of exposure assessment (dose metrics)	Exposures must be reliably measured
Relevance of effects for the evaluation of influences on human health	The relevance of the studies to human health is dependent on two factors: firstly, whether the studies, when not conducted on humans, are transferable to humans, and, secondly, whether the measuring point is relevant to human health. Four gradations are possible: – The effects have an influence on health – The effects plausibly indicate harm – The possibility that the effects are harmful to health can be neither eliminated nor assumed – The effects are within a tolerable range, even for long-term exposure

Step 5: Hazard characterization in the problem areas

Building on this summary, the overall scientific picture must be characterized in the various experimental fields. In this, discrepancies and contradictions are to be clarified as far as possible.

What are central are the arguments that speak for and against the existence of a hazard. These must be thoroughly elaborated.[108] Under the arguments for, all the arguments that indicate a hazard to humans are collected. As arguments against, all the arguments that speak against such a suspicion are considered. This qualitative hazard characterization should serve the citizens' advisory council as an evaluative foundation. Therefore particular attention is to be paid to transparency and comprehensibility here.

A consensus about the arguments for and against is a necessary precondition for the advancement of the EUR, because every evaluator has to weigh these arguments. However, no agreement need be reached. What is far more significant is how the evaluators differ from each other in the appraisal of the hazard.

A quantitative evaluation of the arguments for and against may be conducted in various ways. Firstly, a scale of confidence grades must be formed. The California EMF Project (Neutra et al. 2002) provides an example of this. Seven gradations for the degree of confidence were made, to which numerical values were assigned. These express the degree of confidence on a scale from 0 to 100% confidence (see Table 34).

Table 34 Definitions of confidence grades. (From Neutra et al. (2002, p. 61).)

Qualitative description	Confidence (%)
Virtually certain that electromagnetic fields increase the risk to some degree	>99.5
Strongly believe that they increase the risk to some degree	90–99.5
Prone to believe that they increase the risk to some degree	60–90
Close to the dividing line between believing or not believing that they increase the risk to some degree	40–60
Prone to believe that they do not increase the risk to some degree	10–40
Strongly believe that they do not increase the risk to some degree	0.5–10
Virtually certain that they do not increase the risk to some degree	<0.5

The judgments of confidence indicate how well it is estimated that the evidence for each endpoint is evidence for a causal connection between exposure and adverse effect, and therefore for the existence of a hazard. The judgments of confidence also enable a direct comparison of the experts regarding their appraisal of the evidence. The arithmetic average can be taken here as a summary characteristic.

[108] See the EPA handbook on risk characterization (US EPA 2000).

Step 6: Evaluation by the citizens' advisory council
On the basis of the arguments for and against, the confidence table, and the competence profiles of the experts, the representatives of the citizens' advisory council can then produce a ranking and the hazard evaluated be classified according to urgency. Categories such as "not urgent", "moderately urgent", or "very urgent" can be used for this. An alternative is the classification by various precautionary options.

In order to guard against possible misunderstanding, the citizens' advisory council should not judge any evidence, but should arrive at a considered evaluation of the necessity of action on the basis of the evaluation of evidence by the experts and in full knowledge of the differences between the experts and their reasons (i.e., the respective weightings of the arguments for and against).

Appendix 4:
Comparative Evaluation of the Risks of Hazardous Incidents

In the case of risks associated with hazardous incidents, the uncertainty typical of risks mainly has less to do with whether exposure leads to harm or damage but rather whether an emission takes place. In contrast to normal risks, the probability of the occurrence of a pollutant (e.g., the emission) is to be considered as a further attribute in the characterization of risk. Their scale reaches from events which, though far from everyday, are within the realm of human experience (e.g., hazardous incidents in chemical facilities, airplane crashes) to events which are so improbable that one does not practically anticipate (e.g., airplane crash in an inner city, breach of a dam) and which one can therefore describe as hypothetical.

The conduct of a comparative evaluation of risks of hazardous incidents differs only slightly from the multiattribute comparative evaluation of normal risks. As Table 35 shows, the aim, focus, participants, and output are the same as for normal risks. The difference is that the probability of the occurrence of a pollutant cannot (at least, not generally) be allocated in a linear fashion against the extent of damage, and hence the different probability fields must be separately evaluated. The organization and conduct of a comparative evaluation of risks of hazardous incidents corresponds to the steps of a multiattribute comparative evaluation of normal risks as described in Appendix 2. These will not, therefore, be discussed again here.

In the following, therefore, just a few of the problems specific to risks arising from hazardous incidents will be presented as an example of the comparative evaluation of technical risks.

Table 35 Profile of a comparative evaluation of the risks of hazardous incidents.

Aim	Comparison of various risks of hazardous incidents on one or more attributes by one or more evaluators
Field of application	Risks of hazardous incidents, i.e., those in which the uncertainty typical of risks is whether exposure to a pollutant takes place
Focus	Assessment of health risks
Participants	Experts and, if need be, citizens
Output	Risk ranking

Comparative Risk Assessment. Holger Schütz, Peter M. Wiedemann,
Wilfried Hennings, Johannes Mertens, and Martin Clauberg
Copyright © 2006 WILEY-VCH Verlag GmbH & Co. KGaA, Weinheim
ISBN 3-527-31667-1

The clarification and limitation of the problem field, as well as the selection of the technical systems to be considered, is the first necessary step. In the following, technical risks from the fields of energy, mobility, and chemicals will be considered. This means that risks associated with power generation, transport systems, or chemical production facilities are considered. These technical risks have both normal risk components and hazardous incident risk components.

One graphic possibility for the simultaneous presentation of damage frequency and extent of damage is a damage frequency diagram. This can be a continuous diagram, for example as given in DRS-A (1979). Here data for airplane, mining, and train catastrophes, as well as fires and explosions, are reproduced. These data are based in part on statistical data from Great Britain and partly on an extrapolation of worldwide data (Figure 29).[109]

As Figure 29 shows, no ranking exists which is valid for all fields of frequency. The ranking depends very much more on which section of the damage frequency diagram is considered. For instance, a lower extent of damage arises for fire and explosion accidents, which occur more frequently than once in 100 years than for the other risks presented. In contrast, for events occurring less frequently than once in 500 years, fire and explosion accidents have the largest extent of damage.

This is an instance of a comparison in which quantitative assessments of damage frequencies and damage extents are given. For the field of risks associated with hazardous incidents, this may not always be the case.

Should insufficient, or insufficiently precise, data be available, then one can form categories for each comparable risk field. This formation of categories is established practice for the purpose of the safety-engineering systems of facilities. Thus in, e.g., Sicherheitskriterien für Kernkraftwerke [safety criteria for nuclear power facilities] (1977), a distinction is drawn between:

- Anomalous operational circumstances – these can arise during the lifetime of the facility and must be controlled so that hazardous incidents can be avoided with sufficient reliability.
- Hazardous incidents – systems for their control are to be provided (the German Radiation Protection Ordinance calls for verification that the values of Section 28 Part 3 of this ordinance are to be complied with).
- Accidents – measures are to be provided for the determination and containment of the consequences of accidents.

Categories are also considered in the safety-engineering systems of facilities and processes in the chemical industry. Madjar and von Rohr (1995) distinguish the following procedures:

- *Qualitative approach.* An event is verbally described relative to other events. Absolute reference points and specifications for the bandwidths are lacking (in which field is the term "high" to be classified?), so that no comparison of various processes/specifications is possible.

---

[109] The data upon which this figure is based are probably obsolete, but this has no influence on the principle of the procedure.

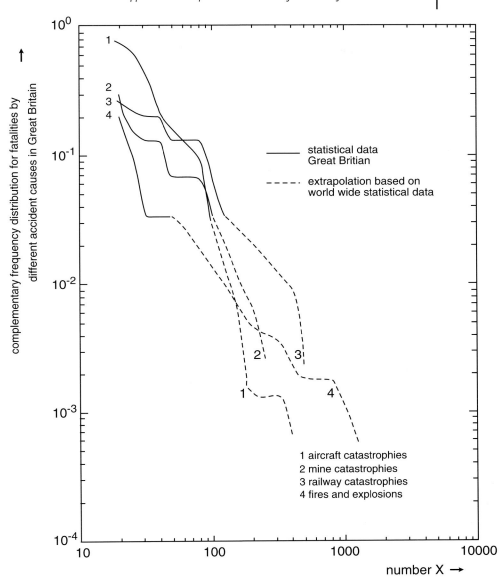

Figure 29
Damage frequency diagram. (From DRS-A (1979).)

- *Semiquantitative approach.* A qualitative term is classified as belonging to a magnitude. The breadth of the classifications should be chosen so that the findings of imprecise data still lie within the bandwidth of the classes.
- *Quantitative approach.* A qualitative term is given a numerical value. The verification of results is made possible by the comprehension of single steps of awareness.

Table 36 provides an example of a semiquantitative risk analysis for chemical facilities. First, a categorization of probabilities (in the sense of relative frequencies) is performed. Table 36 shows the frequency intervals, their verbal descriptions, and their respective probability categories.

Table 36 Example of a semiquantitative evaluation of probability on the process level. (Source Hoffmann-La Roche: Madjar and von Rohr (1995, Table 4.6).)

Probability	Description	
A	Common	More than once a year
B	Frequent	Once a year
C	Occasional	Once in 5 years
D	Rare	Once in 30 years
E	Improbable	Once in 100 years
F	Very improbable	Once in 1000 years

Table 37 contains the categorization of consequences in classes of importance that have been formed from combinations of the consequences for humans, the environment, and material assets. This is simultaneously an example of how a multidimensional evaluative structure (humans, environment, material assets) can be aggregated and categorized around a specific aim. The determination of the contents of these importance classes results from the specific aims of these analyses in the context of chemical safety analyses.

Using the categories from Tables 36 and 37, various hazardous incident scenarios can be placed in a risk matrix according to their probability and importance. Table 38 gives an example of such a matrix with four fictional scenarios.

It can be seen from Table 38 that scenario 4, in importance category II, is by far the most probable. Two scenarios (1 and 3) fall into probability category E. In the comparative risk evaluation of these two risks, scenario 3 will come before scenario 1, because it has the greater expected harmful consequences.

The risk matrix in Table 38 also makes clear a fundamental problem connected with the use of such qualitative categorizations for comparative risk evaluation. When one considers all four scenarios, it becomes clear that not all may be unambiguously ranked. Only scenario 4 lies clearly in rank 1, because it has the highest probability and no other scenario has a higher importance ranking. In contrast, the rank formation will determine the rankings of the other three scenarios. In a lexi-

Table 37 Example of a semiquantitative importance assessment. (Source Hoffmann-La Roche: Madjar and von Rohr (1995, Table 4.2)[a].)

Importance class	Description	
I	Humans:	dead, evacuated out of the area
	Environment:	long-term damage throughout the area
	Material assets:	>10 million sFr, outage of facility > 1 year
II	Humans:	injuries, irritations outside the area
	Environment:	reversible damage in the neighborhood
	Material assets:	<10 million sFr, outage of facility: months
III	Humans:	injuries only in the area, nuisances in the neighborhood
	Environment:	at most the sewage works of the area are affected
	Material assets:	<2 million sFr, outage of facility: weeks
IV	Humans:	slight injuries only in the area
	Environment:	only the area of the facility affected
	Material assets:	<1 million sFr, outage of facility: days

[a] "A semiquantitative classification (Table 4.2) often suffices to assess the order of magnitude of the importance of individual scenarios on the process/facility level. However, should effects of greater importance be ascertained, then the models here for reckoning release and dispersal are necessary."

Table 38 Example of a risk matrix. (Modified from Madjar and von Rohr (1995, Figure 4.6).)

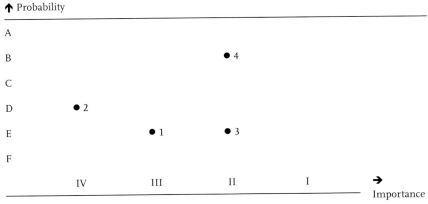

● 1 = scenario number 1.

cographic classification,[110] which this suggests, the ranking depends on which of the two evaluative criteria, probability or importance, is seen as more significant. If one weights probability higher than importance, the ranking of scenarios will be 4, 2, 3, 1. Should one weight importance higher than probability, the ranking 4, 3, 1, 2 arises.

Noncompensatory evaluative procedures provide no opportunity here for further analysis – the ranking is simply the outcome of the relative weighting of probability and importance (and, naturally, the evaluation of the scenarios according to both attributes).

One possibility for a differentiated evaluation on the basis of the given categories is the multiattribute risk evaluation explained in Section 5.2.2 and illustrated in Appendix 2. In this procedure, probability and importance are the attributes by which the four scenarios are evaluated. Both attributes may, according to the estimations of the evaluators, be weighted differently or identically. The scenarios can then be placed by the evaluators on the rating scale according to their probability and importance and are thus amenable to a differentiated appraisal. It is also in principle possible to construct nonlinear utility functions for both attributes that permit a consideration in the evaluation of the great differences in characteristics within the categories of probability and importance.

---

110) The rank formation proceeds such that the scenario with the highest estimation on the most important attribute takes the highest rank. Should several scenarios have the same estimations, then the second most important attribute is considered in the same manner (see Section 5.2.2, the lexicographic rule).

References

Alhakami, A.S. (1991) A psychological study of the inverse relationship between perceived risk and perceived benefit of technological hazards. *Dissertation Abstracts*, Vol. 52/09-B.

Alhakami, A.S. and Slovic, P. (1994) A psychological study of the inverse relationship between perceived risk and perceived benefit. *Risk Analysis* 14, 1085–1096.

Apostolakis, G.E. and Pickett, S.E. (1998) Deliberation: integrating analytical results into environmental decisions involving multiple stakeholders. *Risk Analysis* 18 (5), 621–634.

APUG (1999) *Action Programme Environment and Health*. Federal Ministry for Health and the Federal Environment Ministry, Germany (available at: http://www.apug.de/archiv/pdf/Action_Programme_1999.pdf (accessed 17 Jan. 2006)).

Ayyub, B.M. and Bender, W.J. (1999) *Assessment of the Construction Feasibility of the Mobile Offshore Base. Part I – Risk-Informed Assessment Methodology*. College Park, MD: Center for Technology and Systems Management, University of Maryland.

Bailar III, J.C. and Bailer, A.J. (1999) Risk assessment – the mother of all uncertainties. Disciplinary perspectives on uncertainty in risk assessment. In A.J. Bailer, C. Maltoni, J.C. Bailar III, F. Belpoggi, J.V. Brazier and M. Soffritti (eds.), *Uncertainty in the Risk Assessment of Environmental and Occupational Hazards: An International Workshop*, Vol. 895, pp. 273–285. New York: New York Academy of Sciences.

Balderjahn, I. and Wiedemann, P.M. (1999) *Bedeutung von Risikokriterien bei der Bewertung von Umweltproblemen*. Arbeiten zur Risiko-Kommunikation Heft 74. Programmgruppe Mensch, Umwelt, Technik. Forschungszentrum Jülich.

Baron, J. and Spranca, M. (1997) Protected values. *Organizational Behavior and Human Decision Processes* 70, 1–16.

Bazerman, M.H. (1990) *Judgment in Managerial Decision Making*. New York: John Wiley.

Bazerman, M.H. and Lewicki, R.J. (1983) *Negotiation in Organizations*. Beverly Hills/London/New Delhi: Sage Publications.

Bazerman, M.H. and Neale, M.A. (1993) *Negotiating Rationality*. New York: Free Press.

Bazerman, M.H., Moore, D.A., Tenbrunsel, K.A., Wade-Benzoni, K.A. and Blount, S. (1999) Explaining how preferences change across joint versus separate evaluation. *Journal of Economic Behavior and Organization* 39, 41–58.

Becher, H., Steindorf, K. and Wahrendorf, J. (1995) *Epidemiologische Methoden der Risikoabschätzung für krebserzeugende Umweltstoffe mit Anwendungsbeispielen*. Berichte des Umweltbundesamts, 7/95. UBA-FB 95-041 (Umweltforschungsplan 116 06 089). Deutsches Krebsforschungszentrum, Heidelberg. Berlin: Erich Schmidt Verlag.

Beck, U. (1986) *Risikogesellschaft. Auf dem Weg in eine andere Moderne*. Frankfurt am Main: Suhrkamp.

Bennett, D.H., Margni, M.D., McKone, T.E. et al. (2002) Defining intake fraction. *Environmental Science and Technology* 36 (9), 206A–211A (available at: http://pubs.acs.org/cgi-bin/article.cgi/esthag-a/0000/36/i09/html/09benn.html (accessed 17 Jan. 2006)).

Berg, M., Erdmann, G., Leist, A., Renn, O., Schaber, P., Scheringer, M., Seiler, H. and Wiedenmann, R. (1995) *Risikobewertung im Energiebereich*. Zürich: vdf Hochschulverlag AG an der ETH Zürich.

Beyth-Marom, R. (1982) How probable is probable? A numerical translation of verbal probability expression. *Journal of Forecasting* 1, 257–269.

BfR, Bundesinstitut für Risikobewertung* (2002) *Acrylamid in Lebensmitteln – ernstes Problem oder überschätzte Gefahr?* Informationsveranstaltung des BgVV vom 29.08.2002 (available at: http://www.bfr.bund.de/cm/208/acrylamid_in_lebensmitteln_ernstes_problem_oder_ueberschaetzte_gefahr.pdf (accessed 17 Jan. 2006)). (* Formerly Bundesinstitut für gesundheitlichen Verbraucherschutz und Veterinärmedizin (BgVV)).

Bonano, E.J., Apostolakis, G.E., Salter, P.F., Ghassemi, A. and Jennings, S. (2000) Application of risk assessment and decision analysis to the evaluation, ranking and selection of environmental remediation alternatives. *Journal of Hazardous Materials* 71 (1–3), 35–57.

Bogen, K.T. (1990) *Uncertainty in Environmental Risk Assessment*. New York: Garland Publishing.

Breyer, S. (1993) *Breaking the Vicious Circle: Toward Effective Risk Regulation*. Cambridge, MA: Harvard University Press.

Brun, W. and Teigen, K.H. (1988) Verbal probabilities: ambiguous, context-dependent, or both? *Organizational Behavior and Human Decision Processes* 41, 390–404.

Budescu, D.V. and Wallsten, T.S. (1985) Consistency in interpretation of probabilistic phrases. *Organizational Behavior and Human Decision Processes* 36, 391–405.

Budescu, D.V. and Wallsten, T.S. (1995). Processing linguistic probabilities: general principles and empirical evidence. In J.R. Busemeyer, R. Hastie and D.L. Medin (eds.), *The Psychology of Learning and Motivation: Decision Making from the Perspective of Cognitive Psychology*, pp. 275–318. San Diego, CA: Academic Press.

Büllingen, F., Hillebrand, A. and Wörter, M. (2002) *Elektromagnetische Verträglichkeit zur Umwelt (EMVU) in der öffentlichen Diskussion. Situationsanalyse, Erarbeitung und Bewertung von Strategien unter Berücksichtigung der UMTS-Technologien im Dialog mit dem Bürger*. Studie im Auftrag des Bundesministeriums für Wirtschaft und Technologie (BMWi): WIK Consult, Bad Honnef.

Canadian Standards Association (1996) *A Guide to Public Involvement*, Z764-96. Canada: CSA.

CEC (1996) *Technical Guidance Document in Support of Commission Directive 93/67/EEC on Risk Assessment for New Notified Substances and Commission Regulation (EC) No. 1488/94 on Risk Assessment for Existing Substances.* Luxembourg: Office for Official Publications of the European Communities.

Champaud, C. and Bassano, D. (1987) Argumentative and informative functions of French intensity modifiers "presque" (almost), "à peine" (just, barely) and "à peu près" (about): an experimental study of children and adults. *Cahiers de Psychologie Cognitive/European Bulletin of Cognitive Psychology* 7 (6), 605–631.

Chapman, G.B. and Johnson, E.J. (1995) Preference reversals in monetary and life expectancy evaluations. *Organizational Behavior and Human Decision Processes* 62, 300–317.

Charlton Research Company (2001) *Poll on Comparative Risk.* Congressional Institute.

Chemical Manufacturers Association, CMA (1988) *Title III Community Awareness Workbook.* Washington, DC: CMA.

Clarke, M. and Oxman, A.D. (eds.) (2002) Cochrane Reviewers Handbook 4.1.6 (updated January 2003). In *The Cochrane Library, Issue 2, 2002.* Oxford: Update Software. Updated quarterly (available at: Cochrane Collaboration http:/www.cochrane.org/cochrane/resource.htm).

Clemen, R.T. (1991) *Making Hard Decisions.* Boston, MA: PWS-Kent Publishing.

Cleveland, W.S. (1993) *Visualizing Data.* Murray Hill, NJ: Hobart Press.

Cohen, B.L. (1991) Catalog of risks extended and updated. *Health Physics* 61 (3), 317–335.

Cohen, J.T., Beck, B.D. and Rudel, R. (1997) Life years lost at hazardous waste sites: remediation worker fatalities vs. cancer deaths to nearby residents. *Risk Analysis* 17 (4), 419–425.

Commoner, B. (1992) *Pollution Prevention: Putting Comparative Risk Assessment in Its Place.* Paper presented at the Conference Setting National Environmental Priorities: The EPA Risk-Based Paradigm and Its Alternatives, Annapolis, MD, 15–17 Nov. 1992.

Covello, V.T., Sandman, P.M. and Slovic, P. (1988) *Risk Communication, Risk Statistics, and Risk Comparisons: A Manual for Plant Managers.* Washington, DC: Chemical Manufactures Association.

Crouch, E.A.C. and Wilson, R. (1982) *Risk – Benefit Analysis.* Cambridge, MA: Ballinger Publishing.

Cullen, A.C. and Frey, H.C. (1999) *Probabilistic Techniques in Exposure Assessment: A Handbook for Dealing With Variability and Uncertainty in Models and Inputs.* New York: Plenum Press.

Dienel, P. (2002) *Die Planungszelle.* Der Bürger als Chance. Opladen: Westdeutscher Verlag.

DRS-A (1979) *Deutsche Risikostudie Kernkraftwerke.* Eine Studie der Gesellschaft für Reaktorsicherheit im Auftrag des Bundesministeriums für Forschung und Technologie. Hauptband. Köln: Verlag TÜV Rheinland.

Drüke, H., Henckel, D., Reidenbach, M. and Seidel-Schulze, A. (2003) *Monitoring zur Selbstverpflichtung der Netzbetreiber gegenüber der Bundesregierung Verbesserung der Kooperation mit den Kommunen beim Aufbau von Mobilfunknetzen. Ergebnisse einer Befragung von Kommunen und Netzbetreibern.* Gutachten im Auftrag des Informationszentrums Mobilfunk e.V. (IZMF). Berlin: Deutsches Institut für Urbanistik (DIFU).

EC (2000a) *First Report on the Harmonisation of Risk Assessment Procedures – Part 1* (report). Brussels: European Commission; Scientific Committee's Working Group; Directorate C.

EC (2000b) *First Report on the Harmonisation of Risk Assessment Procedures – Part 2* (report). Brussels: European Commission; Scientific Committee's Working Group; Directorate C.

EC (2001) *Strategy for a future Chemicals Policy.* White Paper (COM(2001) 88 final). Brussels: Commission of the European Communities (available at: http://europa.eu.int/comm/environment/chemicals/pdf/0188_en.pdf (accessed 17 Jan. 2006)).

Edison Electric Institute (1994) *Public Participation Manual.*

Edwards, W. (1977) How to use multiattribute utility measurement for social decisionmaking. *IEEE Transactions on Systems, Man, and Cybernetics* SMC-7, 326–340.

Edwards, W. and Barron, F.H. (1994) SMARTS and SMARTER – improved simple methods for multiattribute utility measurement. *Organizational Behavior and Human Decision Processes* 60, 306–325.

Einhorn, H.J. and Hogarth, R.M. (1978) Confidence in judgment: persistence of the illusion of validity. *Psychological Review* 85, 395–416.

Eisenführ, F. and Weber, M. (1993) *Rationales Entscheiden.* Berlin: Springer-Verlag.

Erev, I. and Cohen, V.L. (1990) Verbal versus numerical probabilities: efficiency, biases, and the preference paradox. *Organizational Behavior and Human Decision Processes* 45, 1–18.

Evans, L., Frick, M.C. and Schwing, R.C. (1990) Is it safer to fly or to drive? *Risk Analysis* 10, 239–246.

ExternE Methodology (1995) *ExternE – Externalities of Energy*, Vol. 2: *Methodology.* Luxembourg: Office for Official Publications of the European Communities (EUR 16521 EN).

ExternE Summary (1995) *ExternE – Externalities of Energy*, Vol. 1: *Summary.* Luxembourg: Office for Official Publications of the European Communities (EUR 16520 EN).

ExternE Methodology Annexes (1997) DGXII ExternE Project. Brussels: European Commission (available at: http://externe.jrc.es/append.pdf (accessed 17 Jan. 2006)).

ExternE Vol. 10: National Implementation (1999) *ExternE – Externalities of Energy.* Luxembourg: Office for Official Publications of the European Communities (EUR 18528).

ExternE Overview (undated) Externalities of Energy. A research project of the European Commission (available at: http://externe.jrc.es/overview.html (accessed 17 Jan. 2006)).

Ezzati, M., Lopez, A.D., Rodgers, A., Vander Hoorn, S. and Murray, C.J. (2002) Selected major risk factors and global and regional burden of disease. *Lancet* 360 (9343), 1347–1360.

Fehr, R. and Mekel, O.C.L. (1999) Probabilistische Expositionsabschätzung in Umweltmedizin und Verbraucherschutz (Workshop) – Präambel. *UWSF – Z. Umweltchem. Ökotox.* 12 (3), 169.

Femers, S. (1993) *Information über technische Risiken: Zur Rolle fehlenden direkten Erfahrbarkeit von Risiken und den Effekten abstrakter und konkreter Informationen.* Frankfurt: Peter Lang Verlag.

Fillenbaum, S., Wallsten, T.S. Cohen, B. and Cox, J. (1991) Some effects of vocabulary and communication task on the understanding and use of vague probability expression. *American Journal of Psychology* 140, 36–60.

Finkel, A.M. (1989) Is risk assessment really too conservative? Revising the revisionists. *Columbia Journal of Environmental Law* 14, 427–468.

Finkel, A.M. and Golding, D. (1994) *Worst Things First? The Debate Over Risk-Based National Environmental Priorities*. Washington, DC: Resources for the Future.

Finucane, M.L., Alhakami, A., Slovic, P. and Johnson, S.M. (2000) The affect heuristic in judgments of risks and benefits. *Journal of Behavioral Decision Making* 13, 1–17.

Fiorino, D.J. (1989) Technical and democratic values in risk analysis. *Risk Analysis* 9, 293–299.

Fischer, K. and Jungermann, H. (1996) Rarely occurring headaches and rarely occurring blindness: is rarely = rarely? *Journal of Behavioral Decision Making* 9 (3), 153–172.

Fischhoff, B., Slovic, P. and Lichtenstein, S. (1978a) Fault trees: sensitivity of estimated failure probabilities to problem representation. *Journal of Experimental Psychology: Human Perception and Performance* 4, 330–344.

Fischhoff, B., Slovic, P., Lichtenstein, S., Read, S. and Combs, B. (1978b) How safe is safe enough? A psychometric study of attitudes toward technological risks and benefits. *Policy Sciences* 9, 127–152.

Fischhoff, B., Watson, S. and Hope, C. (1984) Defining risk. *Policy Sciences* 17, 123–139.

Fisher, R. and Ury, W. (1981) *Getting to Yes: Negotiating Agreement without Giving*. Boston, MA: Houghton Mifflin.

Fiske, A.P. and Tetlock, P.E. (1997) Taboo tradeoffs: reactions to transactions that transgress spheres of exchange. *Journal of Political Psychology* 18, 255–297.

Florida Center for Public Management (1995). *Comparing Florida's Environmental Risks*. Tallahassee, FL: Florida Department of Environmental Protection.

Florig, H.K., Morgan, M.G., Morgan, K.M., Jenni, K.E., Fischhoff, B., Fischbeck, P.S. and DeKay, M.L. (2001) A deliberative method for ranking risks: I. Overview and test bed development. *Risk Analysis* 21 (5), 913–921.

Freudenburg, W.R. and Rursch, J.A. (1994) The risks of "putting the numbers in context". A cautionary tale. *Risk Analysis* 14 (6), 949–958.

Frischknecht, R. et al. (eds.) (1996) *Ökoinventare von Energiesystemen – Grundlagen für den ökologischen Vergleich von Energiesystemen und den Einbezug von Energiesystemen in Ökobilanzen für die Schweiz*, 3rd edn. Zürich: ETHZ/PSI.

Gaido, K.W., Leonard, L.S., Lovell, S., Gould, J.C., Babai, D., Portier, C.J. and McDonnell, D.P. (1997) Evaluation of chemicals with endocrine modulating activity in a yeast-based steroid hormone receptor gene transcription assay. *Toxicology and Applied Pharmacology* 143 (1), 205–212.

GBD (Murray, C.J.L. and Lopez, A.D.) (1996) *The Global Burden of Disease*. World Health Organization, Harvard School of Public Health, World Bank.

GBD (Kay, D., Prüss, A. and Corvalan, C.) (2000) *Methodology of Assessment of Environmental Burden of Disease*. Geneva: World Health Organization.

GBD (Mathers, C.D., Vos, T., Lopez, A.D., Salomon, J. and Ezzati, M.) (2001) *National Burden of Disease Studies: A Practical Guide. Edition 2.0*. Global Program on Evidence for Health Policy. Geneva: World Health Organization.

German Risk Commission (2002) *1. Bericht der Risikokommission*. Salzgitter: Geschäftsstelle der Risikokommission, Bundesamt für Strahlenschutz (available at: http://www.apug.de/archiv/pdf/1.bericht_langfassung.pdf (accessed 17 Jan. 2006)).

German Risk Commission (2003) *Final Report of Risk Commission*. Federal Radiological Protection Agency (available at: http://www.apug.de/archiv/pdf/Risk_Commission_Final_Report.pdf (accessed 17 Jan. 2006)).

Gerrard, M.B. (2000) Risks of hazardous waste sites versus asteroid and comet impacts: accounting for the discrepancies in US resource allocation. *Risk Analysis* 20 (6), 895–904.

Goldstein, B.D. (1990) The problem with the margin of safety: toward the concept of protection. *Risk Analysis* 10 (1), 7–10.

Gonzales, M. and Frenck-Mestre, C. (1993) Determinants of numerical versus verbal probabilities *Acta Psychologica* 83, 33–51.

Graham, J.D. and Wiener, J.B. (1997) Confronting risk tradeoffs. In J.D. Graham and J.B. Wiener (eds.), *Risk versus Risk. Tradeoffs in Protecting Health and the Environment*, pp. 1–41. Cambridge, MA: Harvard University Press.

Gray, G.M. (2000). *Characterizing Risks for Sound Risk Management*. Paper presented at the 3rd International Workshop on Risk Evaluation and Management of Chemicals, Yokohama National University, Yokohama, Japan, 27–28 January 2000 (available at: http://risk.kan.ynu.ac.jp/rmg/Gray.PDF (accessed 17 Jan. 2006)).

Gray, P.C.R. (1996) *Risk Indicators: Types, Criteria, Effects. A Framework for Analysing the Use of Indicators and Comparisons in Risk Communication*. Arbeiten zur Risiko-Kommunikation Heft 56. Programmgruppe Mensch, Umwelt, Technik. Forschungszentrum Jülich.

Gregory, R. and Mendelsohn, R. (1993) Perceived risk, dread and benefits. *Risk Analysis* 13, 259–264.

Gregory, R., Flynn, J. and Slovic, P. (1995) Technological stigma. *American Scientist* 83, 220–223.

Greim, H. and Reuter, U. (2001) Classification of carcinogenic chemicals in the work area by the German MAK Commission: current examples for the new categories. *Toxicology* 166 (1–2), 11–23.

Häfele, W. (1990) Energiesysteme: Eine einführende Problematisierung. In W. Häfele (ed.), *Energiesysteme im Übergang: unter den Bedingungen der Zukunft; Ergebnisse einer Studie des Forschungszentrums Jülich GmbH*, pp. 1–48. Landsberg/Lech: mi-Poller.

Halpern, D.F., Blackman, S. and Salzman, B. (1989) Using statistical risk information to assess oral contraceptive safety. *Applied Cognitive Psychology* 3, 251–260.

Hamm, R.M. (1991) Selection of verbal probabilities: a solution for some problems of verbal probability expression. *Organization Behavior and Human Decision Processes* 45, 1–18.

Hammitt, J.K. (2002) QALYs versus WTP. *Risk Analysis* 22 (5), 985–1001.

Hansen, B.G., van Haelst, A.G., van Leeuwen, K. and van der Zandt, P. (1999) Priority setting for existing chemicals: European Union risk ranking method. *Environmental Toxicology and Chemistry* 18 (4), 772–779.

Harding, C.M. and Eiser, J.R. (1984) Characterizing the perceived risk of some health issues. *Risk Analysis* 4, 131–141.

Hauptmanns, U., Herttrich, M. and Werner, W. (1987) *Technische Risiken. Ermittlung und Beurteilung.* Berlin: Springer.

Heinemeyer, G. (2000) Probabilistische Expositionsabschätzung in Umweltmedizin und Verbraucherschutz: IV. Probabilistische Schätzung der inhalativen Exposition vs. Punktschätzung, dargestellt am Beispiel der Freisetzung von Xylol aus Farben. *UWSF – Z. Umweltchem. Ökotox.* 12 (2), 105–112.

Heinemeyer, G. and Dürkop, J. (2002) Müssen Kinder bei der Risikobewertung besonders berücksichtigt werden? *Umweltmedizinischer Informationsdienst* 1/2002.

Hirschberg, S. and Voss, A. (1999) *Nachhaltigkeit und Energie: Anforderungen der Umwelt.* PSI Proceedings 99-01.

Hirschberg, S., Spiekerman, G. and Dones, R. (1998) *Project GaBE: Comprehensive Assessment of Energy Systems – Severe Accidents in the Energy Sector.* PSI Bericht 98-16. Villigen, Switzerland: Paul Scherrer Institute.

Hobbs, B.F. (1986) What can we learn from experiments in multiobjective decision analysis. *IEEE Transactions on Systems, Man, and Cybernetics* SMC-16, 384–394.

Hofstetter, P. and Hammitt, J.K. (2002) Selecting human health metrics for environmental decision-support tools. *Risk Analysis* 22 (5), 965–983.

Hohenemser, C., Kasperson, R.E. and Kates, R.W. (1985) Causal structure. In: R.W. Kates, C. Hohenemser and J.X. Kasperson (eds.), *Perilous Progress. Managing the Hazards of Technology*, pp. 25–66. Boulder, CO: Westview Press.

Hood, C., Rothstein, H. and Baldwin, R. (2001) *The Government of Risk. Understanding Risk Regulation Regimes.* Oxford: Oxford University Press.

Hornstein, D.T. (1992) Reclaiming environmental law: a normative critique of comparative risk analysis. *Columbia Law Review* 92 (3), 562–633.

Hsee, C. (1996) The evaluability hypothesis: an explanation of preference reversals between joint and separate evaluations of alternatives. *Organizational Behavior and Human Decision Processes* 46, 247–257.

Hubbell, B.J. (2002) *Implementing QALYs in the Analysis of Air Pollution Regulations.* Draft paper. Research Triangle Park, NC: US EPA, Office of Air Quality Planning and Standards, Innovative Strategies and Economics Group (available at: http://www.epa.gov/ttn/ecas/workingpapers/ereqaly.pdf (accessed 17 Jan. 2006)).

ICNIRP (2002) ICNIRP statement: general approach to protection against non-ionizing radiation. *Health Physics* 82 (4), 540–548.

IPCS, International Programme on Chemical Safety (2000). *Human Exposure Assessment.* Geneva: World Health Organization.

Irwin, J.R. (1994) Elicitation rules and incompatible goals. *Behavioral and Brain Sciences* 17, 20–21.

Irwin, J. and Baron, J. (2001) Values and decisions. In S.J. Hoch, H.C. Kunreuther and R.E. Gunther (eds.), *Wharton on Making Decisions*, pp. 131–157. New York: John Wiley.

Jablonowski, M. (1994) Communicating risk. Words or numbers? *Risk Management* 41 (12), 47–50.

Johnson, B.B. (2002) Stability and inoculation of risk comparisons' effects under conflict: replicating and extending the "asbestos jury" study by Slovic et al. *Risk Analysis* 22 (4), 777–788.

Johnson, B.B. (2003a) Are some risk comparisons more effective under conflict? A replication and extension of Roth et al. *Risk Analysis* 23 (4), 767–780.

Johnson, B.B. (2003b) Further notes on public response to uncertainty in risk and science. *Risk Analysis* 23 (4), 781–789.

Johnson, B.B. and Chess, C. (2003) How reassuring are risk comparisons to pollution standards and emission limits? *Risk Analysis* 23 (5), 999–1007.

Johnson, B.B. and Slovic, P. (1995) Presenting uncertainty in health risk assessment: initial studies of its effects on risk perception and trust. *Risk Analysis* 15 (4), 485–494.

Johnson, B.B. and Slovic, P. (1998) Lay views on uncertainty in environmental health risk assessment. *Journal of Risk Research* 1 (4), 261–279.

Jungermann, H. and Slovic, P. (1993) Die Psychologie der Kognition und Evaluation von Risiko. In G. Bechmann (ed.), *Risiko und Gesellschaft*, pp. 167–207. Opladen: Westdeutscher Verlag.

Jungermann, H., Pfister, H.R. and Fischer, K. (1998) *Die Psychologie des Entscheidens*. Heidelberg: Spektrum Akademischer Verlag.

Kaplan, S. (1997) The words of risk analysis. *Risk Analysis* 17, 407–417.

Kaplan, S. and Garrick, B.J. (1981) On the quantitative definition of risk. *Risk Analysis* 1, 11–27.

Karger, C.R. and Wiedemann, P.M. (1998) *Kognitive und affektive Determinanten der intuitiven Bewertung von Umweltrisiken*. Arbeiten zur Risiko-Kommunikation, Heft 64. Programmgruppe Mensch, Umwelt, Technik. Forschungszentrum Jülich.

Keefer, D.L., Kirkwood, C.W. and Corner, J.L. (2003) Perspective on decision analysis applications, 1990–2001. *Decision Analysis* 1 (1), 4–22.

Keeney, R.L. (1980) *Siting Energy Facilities*. New York: Academic Press.

Keeney, R.L. (1992) *Value-Focused Thinking: A Path to Creative Decision Making*. Cambridge, MA: Harvard University Press.

Keeney, R.L. and Raiffa, H. (1976) *Decisions With Multiple Objectives: Preferences and Value Tradeoffs*. New York: John Wiley.

KEMI (2003). *Human Health Risk Assessment* (1/03). Solna, Sweden: Swedish National Chemicals Inspectorate (available at: http://www.kemi.se/upload/Trycksaker/Pdf/Rapporter/Rapport1_03.pdf (accessed 17 Jan. 2006)).

Kincaid, L.E., Meline, J.D. and Davis, G.A. (1995) Chemical and process information: human health hazards summary. In *Cleaner Technologies Substitutes Assessment – A Methodology and Resource Guide*, Chapter 5 (EPA Grant X821-543). Washington, DC: US EPA, Office of Pollution Prevention and Toxics (available at:

http://www.epa.gov/oppt/dfe/pubs/tools/ctsa/index.htm, document source: http://www.epa.gov/oppt/dfe/pubs/tools/ctsa/ch5/mod5-4.pdf (accessed 17 Jan. 2006)).

Konisky, D.M. (1999) *Comparative Risk Projects: A Methodology for Cross-Project Analysis of Human Health Risk Rankings.* Washington, DC: Resources for the Future.

Kreienbrock, L. and Schach, S. (1995) *Epidemiologische Methoden.* Stuttgart: Gustav Fischer Verlag.

Krewitt, W., Mayerhofer, P., Friedrich, R. et al. (1997) *ExternE National Implementation Germany.* Final report, prepared by IER. Stuttgart: Institut für Energiewirtschaft und Rationelle Energieanwendung (IER) (available at: http://externe.jrc.es/ger.pdf (accessed 17 Jan. 2006)).

Kröger, W. (1998) Nachhaltigkeit und Energie – Weniger Beliebigkeit. In: *Nachhaltigkeit – ein Konzept ohne Inhalt?* BATS report 5/98. Basel: Fachstelle BATS (available at: http://www.bats.ch/bats/forum/98nachhaltigkeit_konzept/nachhaltigkeit_konzept.php (accessed 17 Jan. 2006)).

Kröger, W. (2000) Nachhaltigkeit ist messbar. *Energie-Spiegel* 3/Sept. 2000, Projekt GaBE. Villigen, Switzerland: Paul Scherrer Institute (available at: http://gabe.web.psi.ch/energie-spiegel/ (accessed 17 Jan. 2006)).

Kröger, W., Hirschberg, S. and Foskolos, K. (2001) *The Attributes of Sustainability of Energy Options.* MIT CANES Symposium, April 2001.

Kunreuther, H. (2002) Risk analysis and risk management in an uncertain world. *Risk Analysis* 22 (4), 655–664.

Kunreuther, H. and Linnerooth, J. (eds.) (1983) *Risk Analysis and Decision Processes: The Siting of Liquefied Energy Gas Facilities in Four Countries.* Berlin/New York: Springer.

Landrigan, P.J. (1999) Risk assessment for children and other sensitive populations. In A.J. Bailer, C. Maltoni, J.C. Bailar III, F. Belpoggi, J.V. Brazier and M. Soffritti (eds.), *Uncertainty in the Risk Assessment of Environmental and Occupational Hazards: An International Workshop,* Vol. 895, pp. 1–9. New York: New York Academy of Sciences.

LAI (1992) *Krebsrisiko durch Luftverunreinigungen.* Düsseldorf: Länderausschuss für Immissionsschutz. Ministerium für Umwelt, Raumordnung und Landwirtschaft des Landes Nordrhein-Westfalen.

Levitan, L. (1997). *An Overview of Pesticide Impact Assessment Systems a.k.a. Pesticide Risk Indicators* (background paper prepared for the Organization of Economic Cooperation and Development (OECD) Workshop on Pesticide Risk Indicators, Copenhagen, Denmark, 21–23 April 1997). Ithaca, NY: Cornell University Press (available at: http://environmentalrisk.cornell.edu/PRI/Publications/LCL-PestRiskInd7-97.pdf (accessed 17 Jan. 2006)).

Lion, R., Meertens, R.M. and Bot, I. (2002) Priorities in information desire about unknown risks. *Risk Analysis* 22 (4), 765–776.

Lopez, A.D. and Murray, C.C. (1998) The global burden of disease, 1990. *Nature Medicine* 4 (11), 1241–1243.

Lindell, M.K. and Earle, T.C. (1980) *Public Attitudes Toward Risk Tradeoffs in Energy Policy Choices,* final report. Seattle, WA: Battelle Human Affairs Research Centers.

Madjar, M. and von Rohr, P.R. (1995) *Risikoanalyse verfahrenstechnischer Anlagen*. Polyprojekt Risiko und Sicherheit, Leitfäden, no. 5. Zürich: vdf Hochschulverlag AG an der ETH Zürich.

Magat, W.A., Viscusi, W.K. and Huber, J. (1987) Risk-dollar tradeoffs, risk perceptions, and consumer behavior. In W.K. Viscusi and Magat (eds.), *Learning about Risk*, pp. 83–97. Cambridge, MA: Harvard University Press.

Mathers, C.D., Vos, T., Lopez, A.D., et al. (2001). *National Burden of Disease Studies: A Practical Guide*, Edition 2.0. Global Program on Evidence for Health Policy. Geneva: World Health Organization (available at: http://www.who.int/entity/healthinfo/nationalburdenofdiseasemanual.pdf (accessed 17 Jan. 2006)).

Mathers, C.D. et al. (2002) *Global Burden of Disease 2000: Version 2 Methods and Results*. Global Programme on Evidence for Health Policy, Discussion Paper 50. Geneva: World Health Organization (available at: http://www3.who.int/whosis/discussion_papers/zip/paper50.zip (accessed 17 Jan. 2006)).

Maxwell, D.T. (2002) Decision analysis survey: adding insight VI. *OR/MS Today* June.

Mayerhofer, P., Krewitt, W. and Friedrich, R. (eds.) (1997) *Extension of the Accounting Framework*. ExternE Core Project (Maintenance, Improvement, Extension and Application of the ExternE Accounting Framework): final report. Stuttgart: Institut für Energiewirtschaft und Rationelle Energieanwendung (IER) (available at: http://externe.jrc.es/meth2.pdf (accessed 17 Jan. 2006)).

Mazur, A. (1981) *The Dynamics of Technological Controversy*. Washington, DC: Communication Press.

McCloskey, M. (1994) Problems with inappropriately broad use of risk terminology. *Comparative Risk Bulletin* January.

McDaniels, T.L., Gregory, R.S. and Fields, D. (1999) Democratizing risk management: successful public involvement in local water management decisions. *Risk Analysis* 19 (3), 497–510.

Mekel, O.C.L. and Fehr, R. (2000) Probabilistische Expositionsabschätzung in Umweltmedizin und Verbraucherschutz: III. Berücksichtigung von Variabilität und Unsicherheit in quantitativen Risikoabschätzungen (QRA). *UWSF – Z. Umweltchem. Ökotox.* 12 (1), 42–50.

Merkhofer, M.W. (1987). *Decision Science and Social Risk Management*. Dordrecht: Reidel.

Merkhofer, M.W., Conway, R. and Anderson, R.G. (1997) Multiattribute utility analysis as a framework for public participation in siting a hazardous waste management facility. *Environmental Management*, 21 (6): 831–839.

Mertens, J. (1993) Risikoanalyse. In H. Schütz and P.M. Wiedemann (eds.), *Schlüsselbegriffe der Technikbewertung*, pp. 187–192. Frankfurt/Main: IMK.

Minard, R.A. (1993) *Critical Values at Risk: Lessons from Vermont's Quality of Life Analysis*. South Royalton, VT: Northeast Center for Comparative Risk, Vermont Law School.

Minard, R.J. (1994) *A Practitioner's Guide to Comparative Risk – and How We Got Here*. Washington, DC: US EPA (available at: http://www.epa.gov/opperspd/tools6/part1_2.htm (accessed 24 October 2002)).

Morgan, M.G. and Henrion, M. (1990) *Uncertainty. A Guide to Dealing With Uncertainty in Quantitative Risk and Policy Research.* Cambridge: Cambridge University Press.

Morgan, M.G., Florig, H.K., DeKay, M.L. and Fischbeck, P. (2000) Categorizing risks for risk ranking. *Risk Analysis* 20 (1), 49–58.

Morgan, K.M., DeKay, M.L., Fischbeck, P.S. et al. (2001) A deliberative method for ranking risks: II. Evaluation of validity and agreement among risk managers. *Risk Analysis* 21 (5), 923–937.

Morgenstern, R.D. and Sessions, S.L. (1988) Weighing environmental risks: EPA's unfinished business. *Environment* 30 (6), 14–30.

Morgenstern, R.D., Shih, J.-S. and Sessions, S.L. (2000) Comparative risk assessment: an international comparison of methodologies and results. *Journal of Hazardous Materials* 78 (1–3), 19–39.

Mosbach-Schulz, O. (1999) Probabilistische Expositionsabschätzung in Umweltmedizin und Verbraucherschutz: II. Methodische Aspekte probabilistischer Modellierung. *UWSF – Z. Umweltchem. Ökotox.* 11 (5), 292–298.

Moxey, L.M. and Sanford, A.J. (1993) *Communicating Quantities. A Psychological Perspective*, Essays in Cognitive Psychology. Hove, UK: Erlbaum.

Murphy (2001) Framing the nicotine debate: a cultural approach to risk. *Health Communication* 13 (2), 119–140.

Murray, C.J.L. and Lopez, A.D. (eds.) (1996a) *The Global Burden of Disease. A Comprehensive Assessment of Mortality and Disability from Diseases, Injuries and Risk Factors in 1990 and Projected to 2020*, GBD Series Vol. I. Cambridge, MA: Harvard School of Public Health on behalf of the World Health Organization and the World Bank.

Murray, C.J.L. and Lopez, A.D. (1996b) *The Global Burden of Disease and Injury Series, Executive Summary*. Burden of Disease Unit, Harvard School of Public Health (available at: http://www.hsph.harvard.edu/organizations/bdu/GBDseries.html (accessed 17 Jan 2006)).

National Health and Medical Research Council (NHMRC) (1999) *A Guide to the Development, Implementation and Evaluation of Clinical Practice Guidelines.* Canberra: NHMRC.

National Health and Medical Research Council (NHMRC) (2000) *How to Use the Evidence: Assessment and Application of Scientific Evidence.* Canberra: NHMRC (available at: http://www.nhmrc.gov.au/publications/_files/cp69.pdf (accessed 17 Jan 2006)).

National Research Council (1983) *Risk Assessment in the Federal Government: Managing the Process.* Washington, DC: National Academy Press.

National Research Council (1989) *Improving Risk Communication.* Washington, DC: National Academy Press.

National Research Council (1994) *Science and Judgment in Risk Assessment.* Washington, DC: National Academy Press.

National Research Council (1996) *Understanding Risk. Informing Decisions in a Democratic Society.* Washington, DC: National Academic Press.

NEA, Nuclear Energy Agency (2000) *Nuclear Energy in a Sustainable Dvelopment Perspective.* Paris, France: Nuclear Energy Agency.

Neale, M.A. and Bazerman, M.H. (1985) The effects of framing and negotiator overconfidence on bargaining behaviors and outcomes. *Academy of Management Journal* 28 (1), 34–49.

Neubert, D. (1994) Möglichkeiten und Methoden der quantitativen Risikoabschätzung. In H. Marquardt and S.G. Schäfer (eds.), *Lehrbuch der Toxikologie*, pp. 840–913. Mannheim: BI Wissenschaftsverlag.

Neus, H., Sagunski, H., Kappos, A. and Schümann, M. (1995) Zur administrativen Umsetzung von Risikoabschätzungen (Teil I). *Bundesgesundheitsblatt* 7, 258–264.

Neus, H., Ollroge, I., Schmid-Höpfner, S. and Kappos, A. (1998) *Aktionsprogramm Umwelt und Gesundheit, Teilvorhaben: Zur Harmonisierung gesundheitsbezogener Umweltstandards – Probleme und Lösungsansätze*. Forschungsbericht 116 01 001, UBA-Projekt 98/01, UBA-FB 97-095. Behörde für Arbeit, Gesundheit und Soziales, Amt für Gesundheit, Hamburg, im Auftrag des Umweltbundesamts. Berlin: Erich Schmidt.

Neutra, R., DelPizzo, V. and Lee, G.M. (2002) *An Evaluation of the Possible Risks from Electric and Magnetic Fields (EMFs) from Power Lines, Internal Wiring, Electrical Occupations and Appliances*, final report. Oakland, CA: California EMF Project (available at: http://www.dhs.ca.gov/ps/deodc/ehib/emf/RiskEvaluation/riskeval.html (accessed 17 Jan. 2006)).

Newstead, S.E. and Collis, J.M. (1987) Context and the interpretation of quantifiers of frequency. *Ergonomics* 30, 1447–1462.

NHS Centre for Reviews and Dissemination (2001) *Undertaking Systematic Reviews of Research on Effectiveness: CRD's Guidance for Carrying Out or Commissioning Reviews* (available at: http:/www.york.ac.uk/inst/crd/report4.htm (accessed 17 Jan. 2006)).

Nisbett, R.E. and Ross, L. (1980) *Human Inference: Strategies and Shortcomings of Social Judgment*. Englewood Cliffs, NJ: Prentice-Hall.

NJDEP (2003) *Final report of the New Jersey Comparative Risk Project*. Trenton, NJ: New Jersey Department of Environmental Protection (available at: http://www.state.nj.us/dep/dsr/njcrp/ (accessed 17 Jan. 2006)).

OECD (2003) *Workshop on Near Earth Objects: Risks, Policies and Actions*. Frascati, Italy: Organization for Economic Cooperation and Development.

OSPAR Commission (2000) *Briefing Document on the Work of DYNAMEC and the DYNAMEC Mechanism for the Selection and Prioritization of Hazardous Substances*.

OSPAR Commission (2002) *Dynamic Selection and Prioritization Mechanism for Hazardous Substances*.

Pate-Cornell, M.E. (1996) Uncertainties in risk analysis: six levels of treatment. *Reliability Engineering and System Safety* 54 (2–3), 95–111.

Peters, H.P. (1993) Risikokonflikt/Risikokontroverse. In H. Schütz and P.M. Wiedemann (eds.), *Schlüsselbegriffe der Technikbewertung*, pp. 203–208. Frankfurt/Main: IMK.

Powell J.C. (1996) The evaluation of waste management options. *Waste Management and Research* 14, 515–526.

Presidential/Congressional Commission on Risk Assessment and Risk Management (1997) *Framework for Environmental Health Management*, Vols 1 and 2. Washington, DC (available at: http://cfpub.epa.gov/ncea/cfm/pcrarm.cfm?ActType=default (accessed 17 Jan. 2006)).

Pruitt, D. and Rubin, J.Z. (1986) *Social Conflict: Escalation, Impasse, and Resolution*. Reading, MA: Addison-Wesley.

Raiffa, H. (1982) *The Art and Science of Negotiation*. Cambridge, MA: Harvard University Press.

Ramsberg, J.A. and Sjöberg, L. (1997) The cost-effectiveness of lifesaving interventions in Sweden. *Risk Analysis* 17 (4), 467–478.

Rehbinder, E. (1988) Vorsorgeprinzip im Umweltrecht und präventive Umweltpolitik. In E.U. Simonis (ed.), *Präventive Umweltpolitik*, pp. 129–143. Frankfurt am Main: Campus.

Renn, O., Webler, Th. and Wiedemann, P.M. (eds.) (1995) *Fairness and Competence in Citizen Participation. Evaluating Models for Environmental Discourse*. Dordrecht: Kluwer Academic.

Renn, O., Carius, R., Kastenholz, H. and Schulze, M. (2003). *Entwicklung eines mehrstufigen Verfahrens der Risikokommunikation (ERiK). Ein Leitfaden für die Oberen Bundesbehörden*. Stuttgart: Akademie für Technikfolgenabschätzung in Baden-Württemberg (im Auftrag des Umweltbundesamtes).

Ritov, I. and Baron, J. (1999) Protected values and omission bias. *Organizational Behavior and Human Decision Processes* 79, 79–94.

Rodricks, J.V. (1992) *Calculated Risks*. Cambridge: Cambridge University Press.

Rohrmann, B. and Schütz, H. (1993). The evaluation of decision aiding programs. In S. Nagel (ed.), *Computer-aided Decision Analysis*, pp. 5–29. Westport: Greenwood.

Roth, E., Morgan, M.G., Fischhoff, B., Lave, L. and Bostrom, A. (1990) What do we know about making risk comparisons? *Risk Analysis* 10 (3), 375–387.

Russo, J.E. and Schoemaker, P.J.H. (1990) *Decision Traps*, pp. 75–76. New York: Simon and Schuster.

Salomon, J.A. et al. (2001) *Methods for Life Expectancy and Healthy Life Expectancy Uncertainty Analysis*, Global Programme on Evidence for Health Policy, working paper 10. Geneva: World Health Organization (available at: http://www3.who.int/whosis/discussion_papers/zip/paper10.zip (accessed 17 Jan. 2006)).

Schneider, K. (2000) *Anwendungsbeispiele für quantitative Krebsrisikoabschätzungen in der regulatorischen Toxikologie*. Freiburg: Forschungs- und Beratungsinstitut Gefahrstoffe GmbH (FoBiG) (available at: http://www.fobig.de/Downloads/FoBiG-Risk.PDF (accessed 17 Jan. 2006)).

Schneider, K., Konietzka, R. and Schuhmacher-Wolz, U. (2001) Klassierung krebserzeugender Luftschadstoffe für die TA-Luft-Novelle. *Gefahrstoffe – Reinhaltung der Luft* 61 (10), 454–458.

Schoemaker, P.J.H. and Russo, J.E. (2001) Managing frames to make better decisions. In S.J. Hoch, H.C. Kunreuther and R.E. Gunther (eds.), *Wharton on Making Decisions*, pp. 131–157. New York: John Wiley.

Schümann, M. (2000) Probabilistische Expositionsabschätzung in Umweltmedizin und Verbraucherschutz: V. Probabilistische Modelle der Expositionsabschätzung – Möglichkeiten der Validierung und des Vergleichs am Beispiel einer Altlast mit Bodenbelastungen. *UWSF – Z. Umweltchem. Ökotox.* 12 (3), 169–179.

Schütz, H. and Wiedemann, P.M. (2003) Risikowahrnehmung in der Gesellschaft. *Bundesgesundheitsbl-Gesundheitsforsch-Gesundheitsschutz* 46 (7), 549–554.

Schütz, H., Wiedemann, P.M. and Gray, P.C.R. (2000) *Risk Perception Beyond the Psychometric Paradigm.* Arbeiten zur Risiko-Kommunikation, Heft 78. Programmgruppe Mensch, Umwelt, Technik. Forschungszentrum Jülich GmbH.

Schuhmacher-Wolz, U., Konietzka, R. and Schneider, K. (2002) Using carcinogenic potency ranking to assign air contaminants to emission classes. *Regulatory Toxicology and Pharmacology* 36 (3), 221–233.

Shrader-Frechette, K.S. (1995) Comparative risk assessment and the naturalistic fallacy. *Tree* 10 (1), 50.

Sicherheitskriterien für Kernkraftwerke (1977) Der Bundesminister des Innern: *Bekanntmachung von Sicherheitskriterien für Kernkraftwerke.* Bonn: Bundesanzeiger Nr. 206 vom 3.11.1977.

Sielken, R.L. (2000) Risk metrics and cumulative risk assessment methodology for the FQPA. *Regulatory Toxicology and Pharmacology* 31 (3), 300–307.

Silbergeld, E.K. (1995) The risks of comparing risks. *New York University Environmental Law Journal* 3 (2), 405–430.

Sjöberg, L. (1996) A discussion of the limitations of the psychometric and cultural theory approaches to risk perception. *Radiation Protection Dosimetry* 68, 219–225.

Slovic, P., Kraus, N. and Covello, V.T. (1990) What should we know about making risk comparisons. Comment. *Risk Analysis* 10 (3), 389–392.

SRU (1996) *Zur Umsetzung einer dauerhaft umweltgerechten Entwicklung* (Umweltgutachten 1996). Sachverständigenrat für Umweltfragen. Stuttgart: Metzler-Poeschel.

SRU (1999) *Umwelt und Gesundheit – Risiken richtig einschätzen* (Sondergutachten, Dezember 1999). Sachverständigenrat für Umweltfragen. Stuttgart: Metzler-Poeschel.

SSK, Strahlenschutzkommission (1994) *Strahlenschutzgrundsätze zur Begrenzung der Strahlenexposition durch Radon und seine Zerfallsprodukte in Gebäuden,* veröffentlicht im Bundesanzeiger Nr. 155 vom 18 August 1994.

SSK, Strahlenschutzkommission (2001) *Grenzwerte und Vorsorgemaßnahmen zum Schutz der Bevölkerung vor elektromagnetischen Feldern – Empfehlungen der Strahlenschutzkommission.* Bonn: SSK.

Stone, E.R., Yates, J.F. and Parker, A.M. (1994) Risk communication: absolute versus relative expressions of low-probability risks. *Organizational Behavior and Human Decision Processes* 60, 387–408.

Streffer, C., Bücker, J., Cansier, A., Cansier, D., Gethmann, C.F., Guderian, R., Hanekamp, G., Henschler, D., Pöch, G., Rehbinder, E., Renn, O., Slesina, M. and Wuttke, K. (2000) *Umweltstandards: kombinierte Expositionen und ihre Auswirkungen auf den Menschen und seine Umwelt.* Berlin: Springer.

Swartjes, F.A. (2002) *Variation in Calculated Human Exposure. Comparison of Calculations With Seven European Human Exposure Models* (RIVM report 711701030 /2002): Bilthoven, The Netherlands: RIVM, National Institute for Public Health and the Environment.

Tamburlini, G., von Ehrenstein, O.S. and Bertollini, R. (2002) *Children's Health and Environment: A Review of Evidence*. Luxembourg: World Health Organization and European Environment Agency.

Teigen, K.H. (1988) The language of uncertainty. *Acta Psychologica* 68, 27–38.

Teigen, K.H. and Brun, W. (1999) The directionality of verbal probability expressions: effects on decisions, predictions, and probabilistic reasoning. *Organizational Behavior and Human Decision Processes* 80 (2), 155–190.

Teigen, K.H. and Brun, W. (2000) Ambiguous probabilities: when does p = 0.3 reflect a possibility, and when does it express a doubt? *Journal of Behavioral Decision Making* 13, 345–362.

Tengs, T.O., Adams, M.E., Pliskin, J.S., Safran, D.G., Siegel, J.E.,- Weinstein, M.C. and Graham, J.D. (1995) Five-hundred life-saving interventions and their cost-effectiveness. *Risk Analysis* 15 (3), 369–390.

Thalmann, A. (2005) *Risiko Elektrosmog – Wie ist das Wissen in der Grauzone zu kommunizieren?* (Psychologie – Forschung – aktuell, Band 19). Weinheim: Beltz PVU.

Thompson, K.M. and Bloom, D.L. (2000) Communication of risk assessment information to risk managers. *Journal of Risk Research* 3, 333–352.

Thompson, L. (1990) Negotiation behavior and outcomes: empirical evidence and theoretical issues. *Psychological Bulletin* 108 (3), 515–532.

Thompson, L. and Hastie, R. (1990) Social perception in negotiation. *Organizational Behavior and Human Decision Processes* 47, 98–123.

Tufte, E.R. (1983). *The Visual Display of Quantitative Information*. Cheshire, CT: Graphics Press.

US EPA (1987) *Unfinished Business: A Comparative Assessment of Environmental Problems*. Washington, DC: US EPA.

US EPA (1993a) *A Guidebook to Comparing Risks and Setting Environmental Priorities*. Washington, DC: US EPA (available at: http://purl.access.gpo.gov/GPO/LPS383 (accessed 17 Jan. 2006)).

US EPA (1993b) *Reference Dose (RfD): Description and Use in Health Risk Assessments*. Background Document 1A. Washington, DC: US EPA (available at: http://www.epa.gov/iris/rfd.htm (accessed 17 Jan. 2006)).

US EPA (1996) *Characterization of Uncertainties in Risk Assessment with Special Reference to Probabilistic Uncertainty Analysis* (Information Brief EH-413-068/0496). Washington, DC: US EPA.

US EPA (1998a) *Guidelines for Ecological Risk Assessment* (EPA/630/R-95/002F). Washington, DC: US EPA.

US EPA (1998) *Comparative Risk Framework Methodology and Case Study*. SAB Review Draft. Cincinnati, OH: US EPA (available at: http://cfpub.epa.gov/ncea/cfm/recordisplay.cfm?deid=12465 (accessed 17 Jan. 2006)).

US EPA (2000) *Risk Characterization Handbook*. Washington, DC: US EPA.

US EPA (2003) *Draft Final Guidelines for Carcinogen Risk Assessment* (external review draft, February 2003; NCEA-F-0644A). Washington, DC: US EPA Risk Assessment Forum.

US EPA Risk Assessment Forum Technical Panel (2000) *Supplementary Guidance for Conducting Health Risk Assessment of Chemical*

Mixtures (EPA/630/R-00/002): Washington, DC: US EPA (available at: http://www.epa.gov/nceawww1/pdfs/chem_mix/chem_mix_08_2001.pdf (accessed 17 Jan. 2006)).

US EPA Science Advisory Board (1990) *Reducing Risk: Setting Priorities and Strategies for Environmental Protection* (SAB-EC-90-021). Washington, DC: US EPA.

US EPA Science Advisory Board (1999) *An SAB Report on the National Center for Environmental Assessment's Comparative Risk Framework Methodology. A Review by the Drinking Water Committee.* Washington, DC: US EPAScience Advisory Board.

US EPA Science Advisory Board (2000) *Toward Integrated Environmental Decision-Making.* EPA-SAB-EC-00-011.

Uth, H.-J. (1991) Expertise: Risiko-Kommunikation in der Chemie. In H. Jungermann, B. Rohrmann and P.M. Wiedemann (eds.), *Risiko-Kontroversen – Konzepte, Konflikte, Kommunikation.* Berlin: Springer.

Vlek, C. and Stallen, P.J. (1981) Judging risks and benefits in the small and in the large. *Organizational Behavior and Human Decision Processes* 28 (2), 235–271.

Von Winterfeldt, D. and Edwards, W. (1986) *Decision Analysis and Behavioral Research.* Cambridge, MA: Cambridge University Press.

Vose, D. (1996). *Quantitative Risk Analysis.* New York: Wiley.

Wallsten, T.S., Fillenbaum, S. and Cox, J. A. (1986a) Base rate effects on the interpretation of probability and frequency expressions. *Journal of Memory and Language* 25, 571–587.

Wallsten, T.S., Budescu, D.V., Rapoport, A., Zwick, R. and Forsyth, B. (1986b) Measuring the vague meanings of probability terms. *Journal of Experimental Psychology: General* 115 (4), 348–365.

Wallsten, T., Budescu, D.V., Zwick, R. and Kemp, S. M. (1993) Preference and reasons for communicating probabilistic information in numerical or verbal terms. *Bulletin of the Psychonomic Society* 31, 135–138.

WBGU (1998) *Welt im Wandel - Strategien zur Bewältigung globaler Umweltrisiken. Jahresgutachten 1998.* Wissenschaftlicher Beirat Globale Umweltveränderungen. Berlin: Springer.

Weber, E.U. and Hilton, D.J. (1990) Contextual effects in the interpretation of probability words: perceived based rate and severity of events. *Journal of Experimental Psychology: Human Perception and Performance* 16 (4), 781–789.

Weinstein, N.D., Sandman, P.M. and Robert, N.E. (1989) *Communicating Effectively About Risk Magnitudes* (EPA 230/08-89-064). Washington, DC. US EPA, Office of Policy, Planning and Evaluation.

WHO (1994) *Assessing Human Health Risks of Chemicals: Derivation of Guidance Values for Health-Based Exposure Limits* (Environmental Health Criteria 170). Geneva: World Health Organization.

WHO (2000) *Human Exposure Assessment* (Environmental Health Criteria 214). Geneva: World Health Organization.

WHO (2002) *Establishing a Dialogue on Risks from Electromagnetic Fields.* Geneva: Radiation and Environmental Health, Department of Protection of the Human Environment, World Health Organization.

Wiedemann, P.M. (1993) Risikomanagement. In H. Schütz and P.M. Wiedemann (eds.), *Schlüsselbegriffe der Technikbewertung*, pp. 209–213. Frankfurt/Main: IMK.

Wiedemann, P.M. and Balderjahn, I. (1998) *Komparative Risikobewertung und akteursbezogene Konsenschancen*. Arbeiten zur Risiko-Kommunikation, Heft 67. Programmgruppe Mensch, Umwelt, Technik. Forschungszentrum Jülich.

Wiedemann, P.M. and Brüggemann, A. (2001) *Vorsorge aus der Perspektive der Sozialwissenschaft: Probleme, Sachstand und Lösungsansätze*. Arbeiten zur Risiko-Kommunikation Heft 82. Programmgruppe Mensch, Umwelt, Technik. Forschungszentrum Jülich.

Wiedemann, P.M. and Kresser, R. (1997) *Intuitive Risikobewertung – Strategien der Bewertung von Umweltrisiken*. Arbeiten zur Risiko-Kommunikation Heft 62. Programmgruppe Mensch, Umwelt, Technik. Forschungszentrum Jülich.

Wiedemann, P.M., Carius, R., Henschel, C., Kastenholz, H., Nothdurft, W., Ruff, F. and Uth, H.J. (2000) Risikokommunikation für Unternehmen. Düsseldorf: VDI-Verlag.

Wiedemann, P.M., Mertens, J., Schütz, H., Hennings, W. and Kallfass, M. (2001) *Risikopotenziale elektromagnetischer Felder: Bewertungsansätze und Vorsorgeoptionen*. Endbericht für das Bayerische Staatsministerium für Landesentwicklung und Umweltfragen (Arbeiten zur Risikokommunikation, Heft 81). Jülich: Forschungszentrum Jülich GmbH, Programmgruppe Mensch, Umwelt, Technik.

Wiedemann, P.M., Schütz, H. and Thalmann, A.T. (2002a) *Mobilfunk und Gesundheit. Risikobewertung im wissenschaftlichen Dialog*. Programmgruppe Mensch, Umwelt, Technik. Forschungszentrum Jülich.

Wiedemann, P.M., Karger, C.R. and Clauberg, M. (2002b) *Early Risk Detection in Environmental Health*. Teilvorhaben 9. Aktionsprogramm Umwelt und Gesundheit (im Auftrag des Umweltbundesamtes Berlin) Forschungszentrum Jülich, Programmgruppe Mensch, Umwelt, Technik. Umweltforschungsplan des Bundesministeriums für Umwelt, Naturschutz und Reaktorsicherheit, F+E-Vorhaben, 200 61 218/09.

Wiedemann, P.M., Clauberg, M. and Schütz, H. (2003) Understanding amplification of complex risk issues: the risk story model applied to the EMF case. In N. Pidgeon, R. Kasperson and P. Slovic (eds.), *The Social Amplification of Risk*, pp. 286–301. New York: Cambridge University Press.

Williams, P.R. and Paustenbach, D.J. (2002) Risk characterization: principles and practice. *Journal of Toxicology and Environmental Health B* 5 (4), 337–406.

Wilson, G.R. and Crandon, P. (1998) News framing of dioxin as an environmental health risk: when the *New York Times* downplayed the threat. Paper presented at the AEJMC Southeast Colloquium, New Orleans, 13–14 March 1998.

Wilson, R. (1979) Analyzing the daily risks of life. *Technology Review* 81, 40–46.

Wintermeyer, D. (1999) Probabilistische Expositionsabschätzung in Umweltmedizin und Verbraucherschutz: I. Probabilistische Expositionsabschätzung zur Beurteilung der gesundheitlichen Auswirkung von Umweltbelastung. *UWSF – Z. Umweltchem. Ökotox.* 11 (4), 228–233.

Yoon, K.P. and Hwang, C.-L. (1995) *Multiple Attribute Decision Making: An Introduction*. Thousand Oaks, CA: Sage.

Zimmer, A.C. (1983) Verbal versus numerical processing of subjective probabilities. In R.W. Scholz (ed.), *Decision Making Under Uncertainty*. Amsterdam: Elsevier.

Abbreviations of Organizations

ACP	Advisory Committee on Pesticides, UK (http://www.pesticides.gov.uk/acp_home.asp)
APUG	Aktionsprogramm Umwelt und Gesundheit, action program Environment and Health: joint action program of the (German) Ministry for Health and the Ministry for Environment, Nature Conservation, and Nuclear Safety (http://www.apug.de)
BfG	Bundesanstalt für Gewässerkunde, (German) Federal Institute of Hydrology (http://www.bafg.de/)
BfR	Bundesinstitut für Risikobewertung, (German) Federal Institute for Risk Assessment (http://www.bfr.bund.de)
BfS	Bundesamt für Strahlenschutz, (German) Federal Office for Radiation Protection (http://www.bfs.de/)
BGVV	former Bundesinstitut für gesundheitlichen Verbraucherschutz und Veterinärmedizin, (German) Federal Institute for Health Protection of Consumers and Veterinary Medicine, now Bundesinstitut für Risikobewertung (see BfR)
BMU	Bundesministerium für Umwelt, Naturschutz und Reaktorsicherheit, (German) Federal Ministry for Environment, Nature Conservation, and Nuclear Safety (http://www.bmu.de)
BVEL	Bundesministerium für Verbraucherschutz, Ernährung und Landwirtschaft, (German) Federal Ministry of Food, Agriculture, and Consumer Protection (http://www.verbraucherministerium.de/)
CPSC	Consumer Product Safety Commission, USA (http://www.cpsc.gov/)
DIFU	Deutsches Institut für Urbanistik, German Institute of Urban Affairs, Berlin (http://www.difu.de/)
EEA	European Environmental Agency, Copenhagen (http://www.eea.eu.int/)
EPA	Environmental Protection Agency, USA (http://www.epa.gov)
IARC	International Agency for Research on Cancer, Lyon (http://www.iarc.fr/)
LAI	Bund/Länder Arbeitsgemeinschaft für Immissionsschutz, (German) Länder Emission Control Committee (http://www.lai-immissionsschutz.de/)
LAUG	Länderarbeitsgemeinschaft Umweltbezogener Gesundheitsschutz, (German) Länder Committee on Environment Related Health Protection
NAS	National Academy of Sciences, USA (http://www.nationalacademies.org/)
NEA	Nuclear Energy Agency of the OECD, Paris (http://www.nea.fr)
OECD	Organization for Economical Cooperation and Development, Paris (http://www.oecd.org/)

OSHA	Occupational Safety and Health Administration, USA (http://www.osha.gov/)
PSI	Paul Scherrer Institute, Villigen, Switzerland (http://www.psi.ch/)
RKI	Robert Koch Institute, Berlin (http://www.rki.de/)
SAB	Scientific Advisory Board of the EPA
SRU	Rat von Sachverständigen für Umweltfragen, German Advisory Council on the Environment (http://www.umweltrat.de/)
UBA	Umweltbundesamt, (German) Federal Environmental Agency (http://www.umweltbundesamt.de/)
UN/IAEA	United Nations/International Atomic Energy Agency, Vienna (http://www.iaea.org/)
US EPA	See EPA
WBGU	Wissenschaftlicher Beirat der Bundesregierung Globale Umweltveränderungen, German Advisory Council on Global Change (http://www.wbgu.de)
WHO	World Health Organization (http://www.who.int/)

Index

a
acceptability
– of risk 18
– of risk comparisons **80 f**
actors
– see stakeholders
acceptable daily intake (ADI) 156
adverse effect **16**, 111
aggregation of individual evaluations 128
aleatory uncertainty **17**
attributes
– risk attributes 19, **87 ff**, 118 f

b
benchmark dose 156
benefit
– risk-benefit relationship 111 f
biases
– self-serving 103
boundaries
– system boundaries 84

c
classes of knowledge 136
cognitive dissonance 101
communication
– within a CRA process 150
comparative risk evaluation
– methodology **114 ff**
compensatory methods 129
confirmation seeking 99
conflict 98
consequences 13 f, 22, 89, 91, 95, 111
– health consequences 155 ff
consensus
– false consensus effect 100
consistency effect 100
contrast effect 100
control, illusion of 98 f

d
danger **16**, 21 f
decision 7, 21 f, 23 f, 112
decision-making 77 f, 88 ff, **114 f**
disability adjusted life years (DALY) 47 ff, 161 ff
dissent
– between decision-makers 98
– in experts' appraisals 141
dissonance, cognitive 101
dose-effect relationship 17, 59, 107 f, 155 ff
dose-response assessment 107

e
effect
– consistency effect 100
– contrast effect 100
– false consensus effect 100
– framing effect 101
– effect of the first impression 100
egocentricity 103
emotionalization 87
epistemic uncertainty **17**
evidence
– evaluation of evidence 142, 184 ff
expected loss of lifetime (ELL) 160
experts
– risk evaluation by experts vs. lay people 88 ff
exposure assessment 107, 136 f

f
fact
– vs. hypotheticality 137 f
fairness 88, 110, 132
faulty recall 103
first impression
– effect of the f. i. 100
framing 84

– traps   84
– effect   101

h
halo effect   87
hazard   **16**
hazard …
– identification   3, 17, 107, 181, 184
hazardous incidents
– example of CRA for h. i.   189 ff
health
– (definition)   110
health consequences   155 ff
heterogeneous risks
– example of CRA for h. r.   173 ff
hypotheticality
– vs. fact   137 f

i
illusion of control   98 f
incertitude   **17**
indication
– evaluation of evidence   142
individual exposure ratio (IER)   159
intake fraction   160
interest-led perceptions   98

k
knowledge
– classes of knowledge   136

l
lay people
– risk evaluation by lay people vs. experts   88 ff
LOAEL, LOEL   155

m
margin of exposure (MOE)   159
margin of safety (MOS)   159
measures, metrics of risk   **90 ff, 155 ff**
morbidity   160
mortality   160
multi-attribute …
– risk evaluation   **115 ff, 165 ff**
– utility theory   129

n
nature of comparison
– influence on results of comparison   95 f
negativity distortion   100
negotiations
– good negotiations   99

NOAEL, NOEL   155
noncompensatory methods   128

o
optimism, unrealistic   99
overconfidence   84, 98

p
participation
– of stakeholders   145
perceptions
– interest-led   98 f
– risk   19, 82, 86, 140
perspective, interest-led   99
point estimates   19
potential for damage   **16**
precautionary measures   21 f
precautionary principle   22
presentation of information
– effect on risk perception   93
protection from danger   21 f
public risk perceptions, influence of   28

q
qualitative risk factors   14 f, 19, 88
quality adjusted life years (QALY)   44, 161 ff

r
recall, faulty   103
relative risk   159
reference dose   156
risk
– assessment   **17 f**, 139, 141 ff, 168
– assessment of ecological/environmental risks   17 f
– assessment of health risks   17, 107 f
– attributes (dimensions)   19, **87 ff**, 118 f
– categories   **85 f**, 118
– characterization   **17 f**, 107, 169
– communication   **25 f, 78 ff**
– comparison   **20 f, 77 ff**
– concept and definition   **13 ff**
– evaluation   **18 f**, 117, 139, 143
– known / unclear / unknown risk   136
– management   20, **21 f**, 139, 144
– measures, metrics   **90 ff, 155 ff**
– perception   19, 82, 86, 140
– prioritization   20 f
– ranking   169 ff
– regulation   **23 ff**
– risk-benefit relationship   111 f
– stories   86 f

s

scenario method   140
seeking confirmation   99
self-serving biases   103
sensitivity analysis   129 f
severity of an event
– influence on risk perception   94
simultaneous comparison
– vs. sequential comparison   95 f
slope factor   158
stakeholders (actors)   **146 ff**
– multiattribute evaluation with several stakeholders   130 f
stereotyping   99 f
suspicion
– evaluation of evidence   142
system boundaries   84

t

threshold value   17
tolerability of risk   18
trust
– effect on risk perception   81, 94
type of risk comparison   **79**

u

uncertainty   17, 19 f, 62, **105 ff**, **113 f**
– aleatory / epistemic   17
– influence on risk perception   94
– vs. variability   **106 ff**
unclear risks
– comparative evaluation of u. r.   181 ff
unit risk   157
unrealistic optimism   99

v

value judgments   (4)
variability
– vs. uncertainty   **106 ff**

w

weighting functions   123 f
willingness
– to accept (WTA)   162
– to pay (WTP)   162
wishful thinking   99

y

years of life lost (YLL, YOLL)   61, 160 ff

w

worst case estimates   108

z

zero sum assumption   101